DC02 Clear, vacuum cleaner (detail) ✲ JAMES DYSON, 1995.

Banquete chair (detail) ⊛ CAMPANA BROTHERS, 2002.

Dommage à Corbu (detail) ⊛ STEFAN ZWICKY, 1980.

Thonet (detail) ⊗ PABLO REINOSO, 2005.

Ripple chair (detail) ❋ RON ARAD, 2005.

Alda lamp (detail) ® GAETANO PESCE, 2004.

talk about design

ÉLISABETH COUTURIER

Flammarion

CONTENTS

PREFACE

Rounding a corner on the way to my first day of high school, I stopped dead in my tracks in front of a shop window selling furniture I had never seen before: a chaise longue by Le Corbusier, an armchair by Alvar Aalto, a round table with monoshell chairs by Eero Saarinen, a lamp by Gae Aulenti. It was a revelation. I did not know that this type of furniture existed, in metal, leather, and plastic with pure lines, without embellishment or any reference to the past. And I did not know that design existed. At home, my parents had chosen Empire-style furniture. And all my friends lived in antique-style homes. So, in the face of my discovery, I felt like I had entered a new temporality. There was, then, a before and an after. Before what? After what? I did not have the slightest idea. Something had happened. Something had broken. And nobody around me seemed to have realized what was going on!

Another event also deeply changed my understanding of things in the 1980s. I was visiting Berlin and I had just climbed to the top of a watchtower, from which I could see an area of East Berlin beyond the wall that separated the city in two. Everything was different: the cars, the street furniture, the buildings, the clothes. I felt like I was watching the shooting of a film set in the 1950s. I was surprised to discover that time seemed to have stopped a few hundred meters away; and how much the things that make up our immediate environment depend upon the economic means that produce them. Some time later, the editor of a women's magazine asked me for an article on the different meanings of the, as yet, poorly defined word "design" and its results. I interviewed perfume bottle designers, graphic designers in charge of a business's visual identity, and engineers from research departments specializing in automobile brands or in household appliances, but also fashion and furniture designers. I found out that the appearance of my checkbook or of my tube of suncreen mobilized the powers of specialists whose aim was to render the slightest element of everyday life more attractive. A concern sustained by various, rarely impartial, incentives.

During my research, I noticed the complexity of this particular artistic sphere which depends as much on the inspiration of a designer as on the industrial, technological, scientific, sociological, and ideological reality of the society in which it blossoms.

Finally, nobody can escape design. It seeps into all corners of practical life. But few of us fully understand its influence and resources: it models our behavior, defines our tastes, and follows our expectations. This is why I wanted to clarify its approach in this book. To explain how it came into being; to present its different facets; to talk about the key moments in its history and bring to light its main protagonists, whether they be guiding figures or today's celebrities.

If we turn on the time machine, design is a tremendous benchmark as there is no better indicator of fashions, beliefs, and values. We can understand almost everything about an era through the shape of its lines and the materials it uses. However, plunging into the world of design in order to grasp its ramifications and objectives is more than a simple question of narcissistic mirroring. More than a cosmetic and commercial tool, design maintains such a close relationship with reality that it catches onto its slightest quiver. And for this it deserves particular attention. It establishes an interactive and affective relationship with us that we are not always aware of. Knowing about design also means we know more about the nature of the energy that has run through the history of mankind since modern times. This is why I remember suddenly realizing, in front of the shop window, that it was really exciting to live in the present!

ÉLISABETH COUTURIER

DID YOU SAY DESIGN?

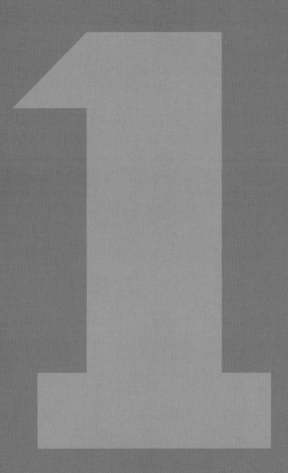

1

WHAT IS DESIGN?

You may think that the objects and furniture created by designers are elitist, expensive, and only relevant to rich yuppies from chic neighborhoods. Think again: we all regularly buy designer objects without even suspecting it. For example, who knew that the Wilkinson Ergonomics plastic razor, along with the Kenwood ST50 iron, was designed by Kenneth Grange, a star of British design? Or that the amusing Firebird gas lighter was dreamed up by the great Guido Venturini? In fact, design works on two levels: on the one hand the "displayed" objects that we admire in the pages of interior decorating magazines; on the other, everyday objects that can be found at the supermarket, department stores, and mobile phone and hi-tech retailers.

But what does the word "design"—used as it is in everyday speech to signify that something has style—mean exactly? It defines objects that most often are mass-produced, conceived, and designed to fulfill various functions—whether to embellish our everyday environment, simplify homemakers' lives, create a brand identity, fight against competition, boost the market, communicate differently, create a fashion, or encourage new behaviors.

The word itself comes from the old French word "desseing." It means both "drawing" (*dessin*) and "intention" (*dessein*), and, if we refer to its distant Latin root, "to represent" and "to designate." **The word plays on this double meaning: to give shape and to testify to a vision of things.** In fact, in concrete terms and according to the *Petit Larousse* dictionary, design is a "discipline aimed at the creation of objects, environments, graphic works, etc., which are func-

tional, aesthetic, and consistent with the requirements of industrial production."

Even more precise, the *Merriam-Webster* dictionary defines design as **"to create, fashion, execute, or construct according to plan."** Design is not therefore simply about the visual aspect of things, it ensures consistency between the technical requirements of fabrication, the object's internal structure, and the way it is used. Related to progress, to the development of ways of thinking, and emerging needs, design gives expression to the economic, social, ideological, and cultural conditions of an era.

Let us move for a moment ten thousand years into the future. Our civilization has disappeared and an archaeologist from another planet discovers several design objects and furniture in the ruins of one of the great Western cities. From detailed observation, he can draw up a precise picture of our economic and social organization, ethics, habits, and tastes. He can thereby deduce our level of industrialization, our scientific progress, our methods of exchange, our use of raw materials, and the state of applied research especially on the subject of materials. He would also be able to understand our domestic habits, life in the city and in the office, and our leisure activities. He could go so far as imagining our discussion of ideas and perhaps even gain an understanding of the morale of our households. The issue of design today is evidently not the same as in the 1940s, an era marked by postwar reconstruction. It is also different from the 1960s, which celebrated the dawn of consumerism. The Western buyer now has a wealth of choice, and functionality is no longer a priority. Design's role is therefore to establish a privileged link with the consumer: appealing in particular to his or her sense of humor, as with Alessandro

Mendini's Anna G corkscrew, which looks like a funny little female statue, or playing with his or her flexibility, as Swatch does by asking him to change his watch as frequently as he changes his shirt.

Design consequently meets our hitherto unconscious desires: satisfying a need, echoing intimate concerns, indicating a social position or a cultural level, appeasing a desire for beauty, or testifying to an interest in the latest technologies. Is this not what Paul Rand, one of the most famous American designers of logos for such renowned companies as UPS, abc, and IBM, meant when he stated that the aim of design was "to transform prose into poetry"?

Sentimental object par excellence, the Teddy Bear Band by Philippe Starck just begs for cuddles (2005).

*FACING PAGE
Firebird, with its masculine forms and vibrant colors, is a domestic gas lighter designed with humor by Guido Venturini (Alessi, 1993).*

Fluid contours for these stylized disposable razors by Kenneth Grange (Schick Silk Effects razor, 1994) and contemporary sobriety for stackable pots and pans by Marc Newson (Tefal, 2003).

FACING PAGE
The stool that never goes out of fashion: Stool 60 by Alvar Aalto was designed in 1932.

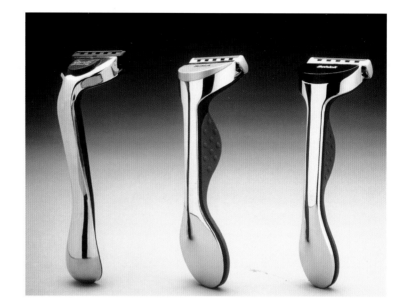

DID YOU SAY DESIGN?

WHEN DID IT START?

VVhether vve appreciate it or not, design is a part of our lives. It touches upon all the areas that make up our day-to-day environment: automobiles, computers, couches, tablevvare, athletic vvear, commercial leaflets, or packs of mineral vvater are conceived, designed, and produced vvith the intention of seducing us. Hovvever, faced vvith a plethora of offers, the tvventy-first century consumer tends to forget that this frenetic beautifying of consumer products, from the most banal to the most sophisticated, is the result of a long history vvhich began about three hundred years ago, and spreads its roots back into the beginnings of the modern vvorld, most notably during the Enlightenment. Inventions such as the steam engine, the loom, and iron architecture appeared in the VVest during this period and contributed a century later to a change in the configuration of the vvorkplace and the structure of society.

By rationalizing the production line in order to reduce the cost of fabrication, the industrial revolution collided head-on with the habits and customs of the so-called "useful arts" (as opposed to the "fine arts" of painting and sculpture), such as carpentry, cabinetmaking, or tapestry vveaving which are based upon manual skills transmitted from one generation to the next. This is why the relationship betvveen crafts and industry became antagonistic, and has been since the very first international exhibition, in London, in 1851. Avvare of the dangers of growing standardization, leading European furniture makers reacted by presenting unique and sumptuous items of furniture in order to shovv the superiority of their skills over the machine. But the appearance of new materials such as steel and wired glass as well as innovative methods such as chemical dyes, or the perfection of complicated techniques such as curved metal and wood, brought about the defeat in the 1920s of this concerted opposition.

Both curious and excited by the stylistic possibilities of these innovations, a handful of forerunners created innovative links between tradition and industry. The opening of the Bauhaus school of applied arts in Germany in 1919, and then the Union des Artistes Modernes (UAM) in Paris in 1929 represent this shift in attitudes. Both initiatives went on to establish fruitful relationships between architects, artists, and engineers. To what purpose? The linking of art and life by connecting the avant-garde aesthetics of, for example, cubism and Russian suprematism, with the possibilities these new techniques offered. And this in order to create a modern, geometrical aesthetic in tune with the times and to get rid of outdated ornamentation and pastiche.

FACING PAGE

This poster of an aerodynamically contoured locomotive against a background of skyscrapers symbolizes American expansion at the end of the 1940s and emphasizes the growing impact of design.

Design as such had not yet been heard of in Europe. **The term makes its official appearance in America after the economic crisis of 1929. Faced with a market on the decline and unable to sell their merchandise, some manufacturers employ for the first time specialists in form such as commercial designers or interior decorators in order to make their products more attractive and more competitive.**

Until now it was taken for granted that form naturally followed function. This is called functionalism. But the involvement of designers, soon to be called industrial designers, showed on the contrary that it was both economically and commercially important to take care of a product's appearance. One of them, a European called Raymond Loewy who spent most of his working life in the United States, wrote the celebrated book *Ugliness Does Not Sell*, whose title aptly sums up the idea. During the 1930s, he invented the celebrated, aerodynamic Streamline, which glorified speed and America's newfound power: automobiles, locomotives, toasters, and gleaming steel lamps symbolizing youth and movement.

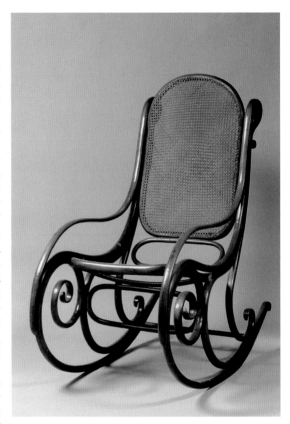

With European reconstruction and the arrival of new materials such as plastic after the war, came the beginning of the golden era of the *Trente Glorieuses* (thirty years of postwar economic properity). For certain brands, design became an efficient means of advertisement: the Olivetti typewriter and the Philips record player introduced a new state of mind in the world of work and leisure. These happy years were dominated by a sense of practicality and a rectilinear aesthetic. The return, at first old hat and humorous then elegant, of curves, patterns, and colors did not happen until the late 1970s. **But for the designer of today, the stakes lie elsewhere. Internet, digital technology, and communication highways represent new challenges: to give form to the elusive and the virtual, and respond to our new style of nomadic living. Design? A constant succession of new beginnings.**

FACING PAGE, *TOP*

This rocking chair,
Fauteuil à Bascule
no.1, is just one
of the many chairs
designed and mass-
produced in Moldavia
by the Thonet
brothers (1860).

FACING PAGE, *BOTTOM*

LC2 Fauteuil Grand
Confort: this armchair,
with geometric lines
and a metal structure,
designed by Le Corbusier,
Pierre Jeanneret,
and Charlotte Perriand
in 1928, is an icon of
the modernist style,
which made its debut
in the 1920s.

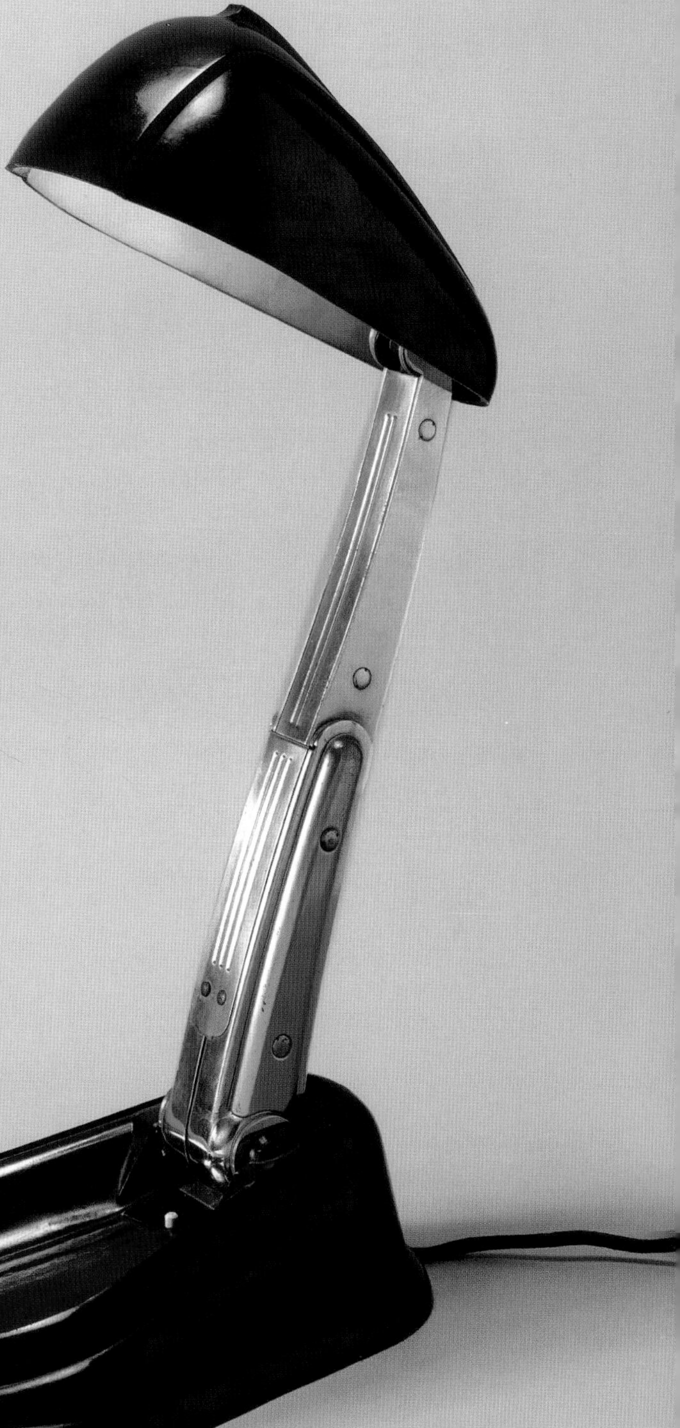

Neither affectation
nor lampshade for
this articulated
Jumo desk lamp in
Bakelite and steel,
made in 1945.

WHERE IS IT FOUND?

Can everything pass through the filter of design? It vvould seem so. Over the past tvventy years, design has filled the slightest chink of our intimate and social lives. Yet, for most of us, it is first and foremost synonymous vvith contemporary furniture. VVhy? Because, vvell before the arrival on the market of the first transistor radios or plastic utensils, furniture designers vvere already concerned vvith questions of form and style. Another common misconception: the belief that design stops at the dravving of a product and that its role is reduced to the embellishment of mundane objects or the reinterpretation of basic forms. Design does indeed belong to a visual universe, and is part of the applied arts. But it is also a vector of services, a communications accelerator, and a revealer of fashion.

As time goes on, these fields of intervention seem unlimited. There are as many types of design as there are areas of reference: product design, graphic design, design of spaces, design of services, etc. In Anglo-Saxon countries, the term design is alvvays accompanied by its specialty: industrial design, fashion design, graphic design, etc. **Today, design applies, among other things, to consumer products, domestic appliances, machine tools, and means of transport. It concerns graphic and typographic creation** with, notably, the invention of new fonts, the conception of a brand's visual identity and logo, the layout of its commercial leaflets, and the visual aspect of its packaging and poster campaigns.

Design also makes its presence felt in street furniture and signs, and the layout of public spaces and commercial outlets. Industry, advertising, the press, publishing, animation, furniture, decoration, and fashion all represent areas of predilection which develop according to technological progress. VVe therefore call upon the talents of a graphic designer to create the image of a web site, make a new computer program more pleasing, or design the characters of a video game. The founders of Apple were the first to involve a forms specialist to make computer access more enjoyable and agreeable for the uninitiated. By establishing an interactive relationship with the user for the first time, the likeable little icons by Susan Kare, who worked for Apple from 1983 to 1986, contributed to making use of the computer easier. A few years later, the iMac's vibrant and transparent envelope by Jonathan Ive made the domestic use of computers even more attractive and contributed to the brand's success.

Sport is another area subject to intense commercial competitiveness and thus on the lookout for powerful identifying signs and codes. Its dogma of surpassing oneself and of high performance rallies the ingenuity of designers. Labels such as Nike and Adidas pay considerable amounts of money in order to own a recognizable visual vocabulary. Three white stripes and the wing of the Greek goddess of victory play an undeniable advertising role. In the same way, skateboarding, snowboarding, surfing, windsurfing, and rollerblading are associated with the use of psychedelic patterns that express the fun spirit of a group of young people drunk on freedom, people who may also dabble with piercing and tattoos, updated by designers of the same generation.

Surprisingly, design can also be found in the kitchens of great chefs. There is nothing more chic than a recipe conceived as much as for the gaze as for the taste buds: the strange coffee splash biscuits made by Radi designers or the "cherry on the cake" designed by Yan

FACING PAGE
To each his logo: The designer Matali Crasset commissioned graphic designers Jean-François Mariceau and Pétra Mzryk to create hers (all round shapes and inspired by her own face). IBM and Paris Match chose the reassuring solidity of a geometrical graphic design.

How to make a technologically sophisticated domestic robot attractive and familiar? By making him look like a toy! (Qrio, Sony, 2003).

Always in fashion, the running shoe fits in with the modern mindset and its emphasis on performance and mobility (Intelligence Level, Adidas, 2005).

D. Pennor for famous French pastry chef Pierre Hermé show that the culinary arts also inspire creators of form. Even more surprising are the last developments in design, such as lighting design, which remodels space with different plays of light, and sound design, which creates musical atmospheres adapted to different places and circumstances. Impalpable and captivating. **Is design always tangible? Not necessarily.**

An original and
amusing version of
Van D. Pennor's
"cherry on the cake"
(a French expression
for "the icing on
the cake") created
for the French
pastry chef
Pierre Hermé:
the cherry breaks
the architectural
lines of this
chocolate cake
(1994).

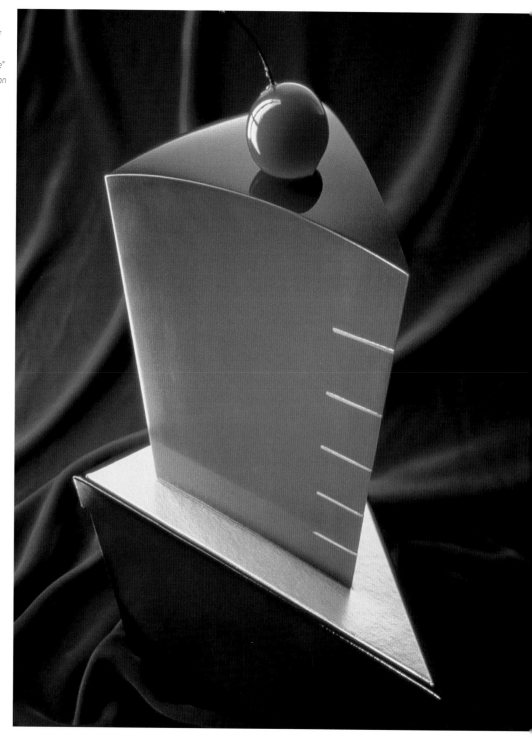

A new interactive friend, this charming Nabaztag rabbit is a little office robot, permanently connected to Wi-Fi, that moves its ears and talks when it gives information. Designed by Christophe Rebours (Violet, 2005).

WHAT DOES TODAY'S DESIGNER DO?

You might think that a designer draws furniture and objects: that his job is simply to make a living-room couch, refrigerator, or toaster beautiful and attractive. Without really knowing much about his work, you imagine him decorating sophisticated interiors or creating amusing gadgets. You are way off target. Today, the term designer covers so many practices that it is difficult for the novice to imagine that one single person could design an automobile, create a table, conceive of a business firm's visual identity, propose musical atmospheres, or imagine perfumed environments.

Neither entirely an artist, nor entirely a technician, the designer hovers between the worlds of industry and art. A photograph by Jean-Baptiste Mondino, inspired by Indian mythology, illustrates with humor the astonishing position of this multi-talented generalist. It shows Philippe Starck dancing like the god Shiva and holding different objects in each of his six hands: a lamp, a shoe, a bottle of mineral water, a toilet brush, a transistor radio, and a paperweight. Inventiveness, flexibility, and adaptability are requisites for carrying out this varied profession. Versed in the tricks of the trade, the designer now enjoys putting on a performance. He leaves nothing to chance. With his black clothes, crew cut, and designer stubble, Starck has created an identifiable look. Others follow his example. Matali Crasset uses her pageboy face for her logo, and the British designer James Dyson himself praises the merits of his first invention the bagless Dyson DC02 vacuum cleaner on huge billboards. Andrée Putman, known as "the diva of design," keeps up her appearance as an elegant Parisian, and non-conformist Ron Arad is always photographed wearing a hat.

The concept of the designer as star has developed since the 1980s. It is a reaction to the growing need to mark out mass-produced objects: a signature gives even the most common object a certain aura. Owning, for example, a Michael Graves whistling kettle, a Naoto Fukasawa CD player distributed by Muji, or a pair of Nike running shoes designed by Marc Newson, transforms the naive consumer into a distinguished connoisseur.

Fashion forerunners, these gurus of form capture unformulated desires, pick up new behaviors, reinvent the ordinary, and foresee new trends. On this subject, Starck explains: "My lemon squeezer is not made for squeezing lemons but for starting conversations."

The designer of the twenty-first century should be able to fulfill all manner of commissions. An acrobatic feat illustrated with humor by Jean-Baptiste Mondino in his portrait of Philippe Starck entitled Diva Shiva, 1998.

As a link between researchers, engineers, businesspeople, and consumers, the designer, according to the visual aspect he gives an object, reflects its technical or technological specificity, its innovation, and the emotional potential it radiates.

The designer must be a multi-tasker who simultaneously has both feet on the ground and his head in the clouds. He must respond to the sometimes contradictory imperatives of a long list of specifications as well as studying the security problems, handling, recycling, and ergonomics of a new product, project, or service. He must chime with the commercial strategy of the firm employing him without betraying his own idea of his social role.

But how does one become a designer?

By graduating from a specialized school, after a four- or five-year program which includes classes in drawing, technology, art, and general culture. And where already the future designer leads projects for various manufacturers. Once graduated, he either begins his profession as part of a team in a big group such as Renault, Braun, or Thomson, or as part of a small agency, or as a freelance designer. His challenge? To adopt an efficient strategy: like Sherlock Holmes, this conceiver of forms,

spaces, and images carries out his investigation. He tries to understand how an object is innovative or irreplaceable, how it provokes interest or rejection, how it changes our ways of living, why it appears at one precise time on the market, and how it can find its place.

Whether he works for public or private companies, the designer has to adapt to a commercial reality, transcend restrictions, represent his own vision of the world, and let his imagination roam. The work of an acrobat!

FACING PAGE

Designers like to be identified with their favorite creations: James Dyson poses with his famous bagless vacuum cleaner, the DC08 (2002), and Marc Newson is photographed sitting on his curious Embryo chair (1988).

Combining solidity and fragility, opacity and transparency, the Gorgonia vase for Daum, designed by Christian Ghion, is imposing in its elegance (2004).

BEFORE DESIGN

2

BEFORE DESIGN

2

BEFORE DESIGN

IN THE BEGINNING
THERE WAS STYLE

A lover of design would never say: "I only like twenty-first century furniture" or "I prefer the postmodern style." Because he is in tune with his time, he would instead explain why he likes the Zeppelin lamp by Marcel Wanders or the Tolozan chaise longue by Eric Jourdan. On the contrary, signature has little or no importance for the buyer of antique furniture: the cabinetmaker's stamp was, according to the period, neither mandatory nor necessary. However, specialists with a practiced eye may be able to ascribe, not without a certain risk, a Louis XIII cabinet to Pierre Golle or, with more certitude, a Louis XIV console to André-Charles Boulle. Before the 1920s in France, several different furniture styles can be discerned; they are named according to the king under whose reign they emerged. From the late Middle Ages to modern times, we can account for twenty or so major styles: Henri II, Louis XIII, Louis XIV, Louis XV, Louis XVI, Regency, Empire, Louis-Philippe, Napoleon, art nouveau, and art deco, to mention only the most famous.

What is style? According to the *Merriam-Webster* dictionary, style is "a distinctive quality, form, or type of something." In fact, we use this term to define the characteristics of forms in a place, during a certain period. Each great style groups together furniture, objects, and a kind of interior design whose aesthetic is recognizable thanks to several clues: the use of straight or curved lines, a specific type of ornamentation, the use of specific materials, etc. A style brings together a group of formal characteristics, an index of distinctive recurring signs and patterns. An enlightened

Black pendants and crystals decorate this chandelier, a sophisticated reinterpretation of the Grand Siècle style, designed by Philippe Starck for Baccarat (2003).

amateur of antique furniture is familiar with these defining symbols. He knows, for example, that the Swedish Gustavian style is defined by the straightness of its fine lines and the whiteness of its lacquered wood, whereas the British country style highlights round forms and warm golden fir wood. If, in literature, style makes a difference, in interior decoration it sums up a period.

WHEN ONE STYLE
DROVE OUT ANOTHER

When we flick through books showing pictures of old, duly stamped interiors, we often feel like we are sitting in front of a stage, after the curtain has lifted and before the actors come on. We enjoy going back in time and imagining different costumes for each period and different ways of moving, sitting, or receiving people. We are transported into sumptuous mansions, in the company of princes and princesses. The history of furniture and decoration generally allows us to relive the great moments of centuries gone by. This is logical. The court, and only the court, set the tone for a long time. The advent of a style depended upon the arrival of a new monarch in power; a necessary but not sufficient condition. The king's desire to leave his mark and instigate important architectural projects, or on the contrary to rely upon the taste of his wife or favorite mistress was also significant.

For the better craftsmen, it was the occasion to show their skills, and to adapt to a change in ways of living. Without a settled population, for example, there were no pleasant and welcoming country mansions, no decorative furniture, no cabinet work by cabinetmakers who came from Holland in the eighteenth century to work on exotic woods. We owe to Louis XIV's lavishness the golden bronze ornaments which adorn the corners and locks of splendid furniture inlaid with copper and tortoiseshell designed by André-Charles Boulle. Likewise, the spacious armchair with widely spaced armrests was invented so that Marie Leszczynska, the wife of Louis XV, could sit comfortably despite her voluminous hoop skirts. Would

A happy coexistence between the French Minister of Culture Jack Lang's gilded office and contemporary furniture designed by Andrée Putman (1984).

the fashion of Egyptian décors and sphinx and scarab ornaments have existed were it not for the impact of the Egyptian campaign that took place just before?

Later, with the advent of modernism, when industrial production had become general practice, another history began: that of design, whose changes in form are, in the end, dictated by the same requirements: pleasing the customer.

FURNITURE AND DECORATION:
a certain expertise

Specialized stores, hypermarkets, and Internet sites now offer a varied choice of furniture, from copies of antiques to the latest designer creations. These products are mostly manufactured in factories, but this has not always been the case. In much of Europe, the making of furniture or decoration was strictly managed by organizations of craftsmen, known as guilds. This model was based on the master-apprentice system and existed to supervise the strict application of regulations governing the craft professions. In return, corporations introduced a system of help and solidarity between specialists, thereby guaranteeing a solid training ground for apprentices. In most cases, only one apprentice was allowed per master to avoid an excess of craftsmen in any one trade.

BEFORE DESIGN

Thus, for centuries, the making of an armchair, for example, was highly codified. Several different professions had to be called upon, in a precise order, for its creation. Firstly, the joiner: he was responsible for the structure, that is to say the legs, the rungs, and the back. The turner then enhanced this with inventive figures: molds, balls, twists, etc. The cabinet maker-sculptor, also called the "image cutter," sculpted inlaid patterns: shells, acanthus leaves, caryatids, and cherubs. The upholsterer then stuffed the arms and back. For ceremonial furniture, the gold painters and bronze makers applied the final touch. Thanks to this incredible organization of complementary talents, French cabinetmaking in the eighteenth century reached a matchless perfection that was recognized throughout the courts of Europe.

Both raw material and finished object, the Cinderella desk designed by Jeroen Verhoeven bridges the gap between the skills of yesteryear and the creations of today (2004).

As for decoration, the building of mansions as early as the fifteenth century demanded the intervention of a coordinator to guarantee stylistic coherence and harmony of line. This was the role of the decorators like the one that François I brought from Italy for the decoration of his château in Fontainebleau. These ancestors of design, who took their ideas from antiquity, nature, or the fashion of the time, were responsible for creating new models. As both designers of furniture and interior decorators, they became the main initiators of new forms: designing columns and balusters, table legs and chairs, and paneling and coffering. They provided their men with engraved plates. Also sold by stamp merchants, these plates inspired carpenters, cabinetmakers, and groups of journeying apprentice craftsmen traveling around France, to improve their technique.

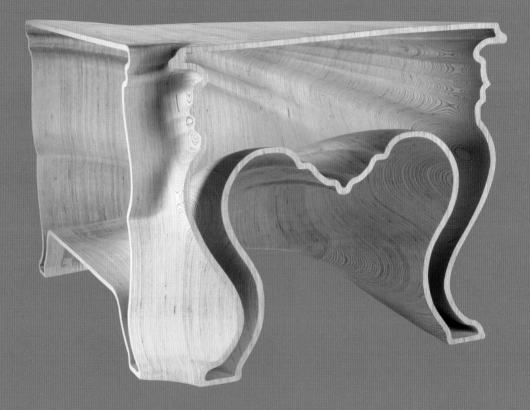

COMMISSIONS OR HOW
TO ESTABLISH AUTHORITY

How did people live in the past? Paintings, even more than drawings or engravings, are irreplaceable testimonies. Quentin Metsys' painting *The Moneylender and his Wife* for example, dated 1514, shows in detail the environment, elements, and objects necessary to a banker of this period in Antwerp, a busy trading capital at the time. Paintings by Veronese reveal the sumptuousness of fifteenth-century Venetian palaces, while Le Nain shows us the poor living conditions of French peasants in the seventeenth century.

Portraits of kings, queens, or nobles in the privacy of their homes provide another kind of information. Paintings such as Quentin de la Tour's portrait of the Marquise de Pompadour illustrate a clever manipulation of appearances, and attempt to overplay the role of these commissioners in the sponsorship of beautiful decorations, fabrics, and paintings. Each detail counts: the shimmering hangings, the delicate ornaments, the refined objects, and the elegant cabinetwork. These compositions testify to a kind of propaganda. For a long time, only the king and those closest to him inspired any changes in furniture or decoration. At the origin of any new style was a fabulous royal commission capable of rallying the best of the profession. François I's project for the building of the châteaux of the Loire valley united French and Italian artists and master builders, permanently marking the fifteenth-century aesthetic. In England, Henry VIII was responsible for the spread of the Italian Renaissance style. Hampton Court Palace in Middlesex and the Palace of Westminster in London both benefited from the talents of Italian craftsmen, featuring the distinctive arabesques, medal-

The armchair is an ageless and universal symbol of power. Alessandro Mendini's Poltrona di Proust offers an exuberant version (1978).

lion heads, and amorini (small winged cherubs) that characterize the style.

Versailles, with its Hall of Mirrors and gilded furniture, did much for the country's glory. If Louis XV's favorite mistress, the Marquise de Pompadour, had not passed her time buying mansions and decorating them with the aid of the best craftsmen and artists of the time, traditional furniture might not have the same reputation. Her brother, the Marquis de Marigny, director of the king's buildings, promoted the Greek style of furniture in fashion around 1755, and the abandonment of the rococo style's sinuous curves. Later, Marie-Antoinette softened the style by feminizing the decoration of the Petit Trianon palace in Versailles. In eighteenth-century America, immigrant craftsmen benefited from wealthy patrons eager to own furniture inspired by Europe's nobility. Limited access to European materials meant that fashions changed more slowly in the colonies, and were available only to the very wealthy.

However, from the Enlightenment onward, the rich bourgeoisie had a hold over commissions that it never relinquished.

ANTIQUES: history is always fashionable

As the year 2000 approached, science-fiction films predicted either a primitive, apocalyptic, and baroque world or, on the contrary, a sterilized and clinical world, in the most pure minimalist style. It must be admitted that these fantastic predictions do not resemble the world we know in any way: the colonization of planets, teleportation, or food in tubes have yet to become a part of our routine. We continue, from one generation to the next, to live with traditional furniture, and we maintain a taste for "beautiful things" when we choose the Regency wing chair, the Louis XV chest of drawers, or the Louis XVI armchair. The antique is always a safe investment for all social classes—and copies sell like hotcakes.

The prestigious antiques fairs in Maastricht, New York, and Paris, the flea markets of London and Prague, and important salesrooms attract a large public of connoisseurs and amateurs, with record prices at stake when a certified, genuine piece comes up. In the history of furniture, styles follow on without necessarily canceling each other out. And the arrival in the 1930s of a pure, rectilinear style changed none of this. Today, it is even considered good taste to mix the ultra contemporary with the antique. Consequently, current designers revisit forms from the past without any qualms. They take from traditional forms indifferently, turning them around, or reinterpreting them. For great styles are part of cultural heritage and will always, therefore, have a real attraction: they carry within them the notion of perfected craftwork and benefit from wide recognition.

Two in one: an example of how to nod in the direction of the past and to contemporary rationality at one and the same time (Silent Chest of Seven Drawers, Becheau-Bourgeois, 2000).

Antique furniture also has other qualities: it is made with traditional, reassuring, and comfortable materials, particularly wood, it testifies to a prestigious past that we would like to have been a part of. Moreover, the antique guarantees the owner's good taste. In a world where everything is speeding up, antique styles have the advantage of their extraordinary permanence. However, 1930s historical, modern design has already joined the ranks of pieces sought after by dealers. The design of pioneers has joined the antiquities department. A ticket for eternity!

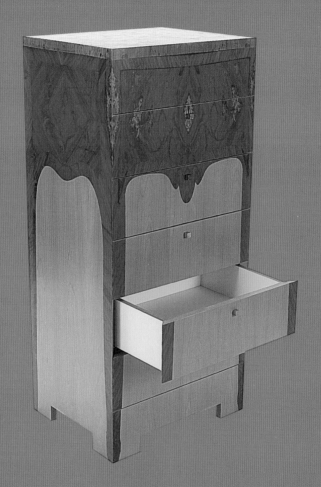

DESIGN AND COMPANY

DESIGN AND TECHNOLOGY:
never one without the other

In the 1960s, the Canadian sociologist Marshall McLuhan declared: "We shape our tools and then our tools shape us." He did not know how right he was. Whether he works with furniture or with industry, today's designer closely follows technological developments and the possibilities offered by new materials. Specialized organizations regularly inform him of the latest developments, especially in areas where research is particularly active: the aerospace industry, biotechnology, and computer science. The history of design has therefore always been closely linked to scientific discoveries, the development of manufacturing processes, and the emergence of new materials. Today, with the Internet, integrated miniature circuits, and "intelligent" materials, the relationship between form and content has considerably changed, leaving more and more room for the imagination, both in terms of the appearance of objects and their method of use.

This was not always the case, however, and the relationship between style creators, industry, and research has not always been an easy one. The case of the Thonet brothers, who, in 1851, created chairs and armchairs with novel contours made according to the mechanized technique of curved wood, was for a long time an exception. They did, however, show that industry and creation could be compatible, and that the union of the two had the advantage of offering the public sophisticated products at reduced cost. **In the end, modernity infiltrated everyday life thanks to the collaboration of** **engineers and architects. Fired up by the first skyscrapers being built in the United States and by the appearance of materials such as steel, reinforced concrete, aluminum, and linoleum, architects began building houses, around the 1930s, with interiors that were completely different from the decoration of previous eras. Ornaments and highly decorated patterns gave way to bright, geometrical forms.** The example of the Maison de Verre (Glass House) built in Paris in 1931 by Pierre Chareau, or of the famous Bauhaus school of applied arts

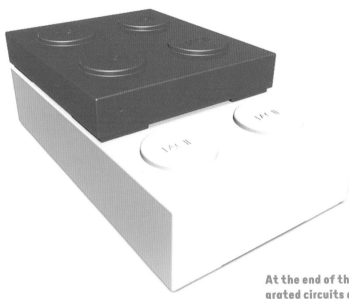

FACING PAGE
The Bic Cristal pen
was born in 1950
and has not gone
out of fashion since—
over a half century
of success already!

Appearances can
be deceptive. This is
not a piece of Lego
but a computer
hard disk (Ora-ito,
LaCie, 2005).

in Dessau, Germany, whose ultra-functional spaces were designed by Walter Gropius down to the last detail a few years before-hand, prove that the concept of rationality had begun to make headway. The fusion of the cabinetmaking workshop and the metals workshop at the Bauhaus school in 1925 marked a revolution embodied by the creation of the first tubular steel chairs signed by Marcel Breuer and Mart Stam. **On the other side of the Atlantic, the dynamism of the New World went hand in hand with an avalanche of inventions and new manufacturing procedures: electrolyzed aluminum, plywood, laminated wood, and fiberglass allowed designers to create pioneering ergonomic forms.** Army research increased innovation during the Second World War. At the end of the war, domestic life was invaded by plastic, Plexiglas, and foam rubber: the Tupperware container with its sealed lid, which allowed for better food conservation, or the Bic Cristal ballpoint pen, a modern and democratic fountain pen, slowly became commonplace accessories.

At the end of the 1950s, the first integrated circuits and the microchip were behind the very first small-format models: radio transistors made their appearance, introduced by the Japanese company Sony during the crazy years of rock and roll. Today, ultra-small wireless appliances, capable of containing tens of thousands of songs, fit into the pocket. **The digital revolution also changed the designer's practice. He now has a new gadget at his disposal: the CAD, or computer-aided design program, which has been used in aerospace and the automobile industry since the 1950s.** Also used by architects, this program can create realistic three-dimensional prototypes. As yet unmade objects can be seen at different angles: transversally, in close-up, or panoramically. This method of virtual creation has given young designers ideas. Ora-ito made his reputation by creating a collection of virtual objects inspired by fashion labels and visible only on the computer screen. Computer-aided manufacturing (CAM) makes the task easier: "We put an idea in at one end, and the product comes out the other end, ready to be packaged, transported, sold, invoiced, and stocked," as the design historian Mel Bryars brilliantly put it (*Design, carrefour des arts*, Flammarion, 2003). This new configuration allows the designer, should he wish, to introduce willful breaks into the program in order to make a series of unique objects.

On the subject of materials, carbon fiber and high-performance industrial ceramics have grown in popularity over the past ten years. The first is used for making racing bikes, golf clubs, or very light and resistant ski helmets. The latter is used for medical aids such as hip replacements or knife blades that never get dull. Another subject of research is convergence. What does this involve? Connecting different technologies together in

order to facilitate ways of using sophisticated materials. Thus avoiding a nervous breakdown on the part of the user! Siemens, for example, offers a tri-band mobile telephone in the form of a pen which is capable of translating hand-written words into a text message, while the likable Nabaztag rabbit (see page 25), the first communicating object, connects itself to the nearest Wi-Fi network: it gives the weather forecast and air quality, can read messages received on voicemail, and so on. Its playful aspect means we forget the ultra-sophisticated technology inside. Which is just as well.

BELOW
With the green movement, the bicycle inspires designers. This one is by Marc Newson (Biomega, 1999).

FACING PAGE
Today's Tupperware adapts to the busy lives of housewives. It stores food and allows it to be reheated in the microwave (2005).

DESIGN AND BUSINESS:
a marriage of convenience

VVho has never experienced this? Sometimes, in the most unexpected of places, at the ends of the earth, vve come face to face with the Michelin man or a huge billboard bearing the red and vvhite colors of Coca-Cola. In Shanghai, Moscovv, and Nevv Delhi, the famous McDonald's yellovv M has become a universal and recurring sign in the urban landscape. VVhen, in 1962, Andy VVarhol painted a can of Campbell's soup, he vvas paying homage to the visual povver of the label, created in 1848 and knovvn to millions of American consumers. Star labels exist vvhose logo and graphic design novv belong to vvorldvvide popular culture. VVhat industrialist has not dreamed, one day, of such publicity?

Next Mother's Day or Father's Day don't overlook the work of Moulinex with the young and fashionable Radi Designers (Principio range of products, 2005).

With the global market and a rapid turnover of products, it has become vital for labels to be immediately identifiable. Numerous studies have shown that a hypermarket customer is confronted by several tens of thousands of products in the space of thirty-five to forty-five minutes. And that he finds his way thanks to labels and packaging. In highly competitive sectors such as food, appliances, automobiles, computers, and athletic equipment, the participation of a designer makes a difference.

The idea of building a brand's identity took off in the United States during the 1950s. The example of IBM who commissioned Eliot Noyes to design the logo and the architecture of its buildings is a reference. It codified everything: colors, graphic design, interior decoration, etc. But the "Big Blue," as the firm was called, was in the end only following in the footsteps of the German company AEG which, in 1909, entrusted its communication to the architect Peter Behrens. This was the first experience of global design, even if the term was only employed later. Behrens participated in the graphic design of the labels, the packaging, and marketing tools, as well as in the signage, and the interior and exterior decoration of the offices. On the strength of this historical precedent, German industrial design came into its own after the Second World War, due to the close collaboration between the Ulm applied arts school and the Braun factory. The sober aesthetic of electric razors, radios, and record players produced by the Frankfurt-based firm—particularly during the 1970s—is the quintessential embodiment of functional, modern, and elegant design. Manufacturers in Italy began to believe in the added value of an attractive object after the Second World War. In addition to the famous Olivetti saga, there is the remarkable one of the steel kitchen utensils and cutlery made by Alessi. Created over the past thirty years by great designers such as Aldo Rossi, the Campana brothers, and Andrea Branzi, salad bowls, table mats, and fruit bowls are as popular with young couples as with collectors. In France, there is the Tim Thom unit by Thomson Multimedia, directed at first by Philippe Starck, which functions as an experimental unit for audiovisual equipment. In the decorative arts tradition, the Ricard firm has since 1995 invited renowned designers such as Olivier Gagnère and Martin Szekely to design their collection of glasses and decanters. And, in 2005, Moulinex entrusted its range of electrical appliances to the Radi Designers group.

According to the sociologist Naomi Klein, the concept of a brand's image has considerably grown today. It has gone beyond the simple visual enhancement of a product to convey lifestyle scenarios with which most people can identify. This is what is called "branding." The slogan of Wrigley's Doublemint Gum is "Double Your Pleasure, Double Your Fun." Conjuring images of freedom, sensuality, and escape, it is just one illustration of this concept. An idyllic vision which the product's design tries to represent in all aspects.

This glass, designed by Martin Szekely, celebrates Perrier (1996).

PAGE 49
The structure
of Citroën's
showroom facade
on the avenue des
Champs-Élysées
in Paris is built
around the famous
chevrons that make
up the firm's logo.
(Architect Manuelle
Gautrand, facade
cover designed by
Gartner, 2006).

Ultra-flashy design
for a good vintage:
champagne bucket
designed by
Marc Newson
(Dom Pérignon,
2006).

DESIGN AND FASHION:
first cousins

Coco Chanel, who had a feel for expressions, once said: "Fashion goes out of fashion, style never does." A frequenter of the artistic avant-garde during the 1920s, she was convinced that sobriety was the height of elegance. She hated frills and fuss. She therefore chose the pure lines of art deco for the decoration of her house, the graphic design of her name, and the form of the Chanel No. 5 perfume bottle. Her rival, the fashion designer Paul Poiret, took a radical line by asking one of the greatest modern architects, Robert Mallet-Stevens, to build and decorate his villa at Mézy-sur-Seine, in the suburbs of Paris. Design and fashion often come together in favor of a shared aesthetic vision, at the crossroads of fashion, or for a one-off collaboration.

In Paris, during the 1980s, in the middle of the baroque fashion movement called Les Nouveaux Barbares, Christian Lacroix asked designers Élisabeth Garouste and Mattia Bonetti to decorate the rooms of his fashion house. It was a success: their heterogeneous style fitted perfectly with the fashion designer's sunny and colorful world. During the same period, a breath of freedom encouraged the de-compartmentalization of disciplines and led to unexpected role play. Thus, Jean-Paul Gaultier transformed himself into a cabinetmaker and made a collection of humorous, hi-tech furniture pieces on wheels conjuring up the idea of traveling, for Valorisation de l'Innovation dans l'Ameublement (VIA). Jean-Charles de Castelbajac who trained at the École des Beaux-Arts in Paris and joined, in 1972, the group Créateurs et Industriels

Avant-garde and geometrical, Hussein Chalayan's clothes conjure up a futuristic appearance (After Words collection, Fall/Winter 2000–01).

founded by Didier Grumbach and Andrée Putman, was one of the first to consider design as a continuation of his career as a fashion designer and to bring about deviations in furniture design. Creating more than simple prototypes, he signed several pieces of furniture including the My Fair Lady chair and the Proust table for the Ligne Roset firm. He also designed brightly colored rugs with geometrical designs. Other fashion designers, particularly the Japanese, went beyond such projects to play with the relationship between the body and clothes. "The body," said Yohji Yamamoto, "is change and transformation." Inspired by his own cultural codes, this master of dissymmetry invented a look that completely broke with traditional French fashion. At the request of Adidas, he designed a collection of urban running shoes combining

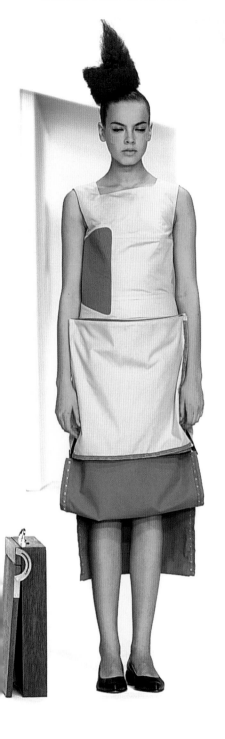

performance and elegance. His "boxing shoes" claimed fashion victims' hearts. Made up of superimposed colored pleats, the Pleats Please collection by Issey Miyake owes its success to its simple cut, the permanent-pleated polyester jersey of which the clothes are made, and the absence of stitching. Minimalist dresses, jumpers, jackets, or pants shape the silhouette. Passionately interested in textile research, the fashion designer opened the Miyake Design Studio in Japan, an experimental laboratory where he still makes new materials by associating industrial creation and craft techniques, technological innovations and old fabrics. Martin Margiela, from northern France and a graduate of the Antwerp school of fine arts, followed in

Jean-Paul Gaultier's footsteps. He charmed the art world by working with secondhand clothes found at flea markets and creating an "unfinished" style. A pure product of the famous Central Saint Martins College of Art and Design in London, the fashion designer Hussein Chalayan declares: "I am essentially a designer, someone who works with ideas." Indeed his creations, using innovative textiles and mate-rials, are closer to theater costumes than func-tional clothing. They follow a very structured, futuristic silhouette where body and clothing overlap to form strange mutating profiles. Less chic than Chanel, but much wilder!

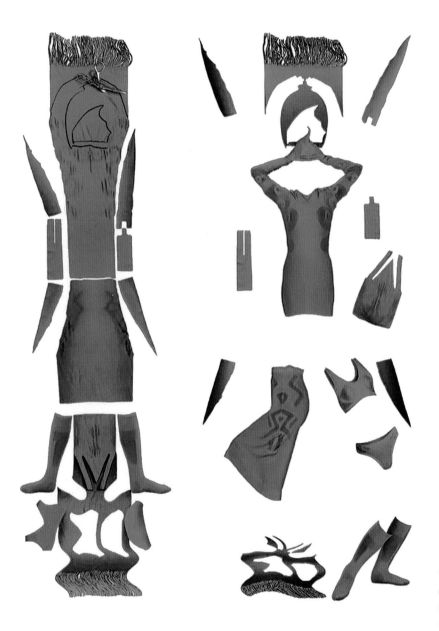
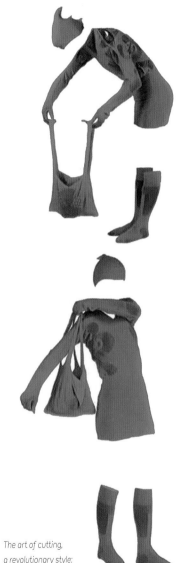

The art of cutting, a revolutionary style: get your scissors ready! (A-POC by Issey Miyake, Miyake Design Studio, 1998).

DESIGN AND CONTEMPORARY ART: an intense flirtation

In 1909, it seemed obvious to Walter Gropius, the founder of the Bauhaus school of applied arts, that he should offer his students art history lessons, as he was convinced that artists were the first to create new forms. Marcel Duchamp, for his part, while visiting the Salon de l'Aéronautique in Paris in 1912, exclaimed to Fernand Léger and Constantin Brancusi who accompanied him: "Painting is finished! Who could do better than this propeller?" Did the future father of the readymade foresee, that day, the place that manufactured objects would come to fill in the modern imagination? In any event, these two apparently contradictory positions show how much and how early the relationship between art and design was considered to be a potentially enriching one.

Certainly, design's functional aspect still marks the boundary separating the applied arts from the fine arts, to the benefit of the latter, which are free of all constraints. But at a time of questions and redefinitions, the line between the two territories is not so clear. **Since the 1960s, following on from the Dadaists' irreverent way with objects in the early twentieth century, many and varied influences have muddied the waters: certain artists flirt with functionality whereas many designers attempt to redefine functional aesthetics.**

During the famous pop art years, artists looked for inspiration in the shelves of supermarket stores. They made poetic combinations with cheap, colorful plastic objects. At the same time, designers, influenced by the free morals which gave rise to events such as those of 1968 in France and motivated by the futuristic visions inspired by space travel, daringly reinvented the domestic environment with their radical proposals. Marc Held designed the Culbuto armchair, the Mourgue brothers designed a cabin, and Verner Panton a Visionna living room which invites laziness and where the floor, walls, and ceiling are joined into one. **The expression "anti-design," claimed in the late 1970s by designers from the Italian movements Superstudio and Archizoom Associati, refers to contemporary design's decision to stand back from modern functionalism.**

This joyful rebellion is at the origin of works often produced in small quantities, some of which are linked to the avant-garde arte

povera movement and its evocation of nature. Sassi and Sodelsasso's giant stones come to mind, or Gruppo Strum's equally large foam grass lawn, or Giulio Paolini's chairs in the form of truncated Ionic columns. In the 1980s, the Memphis and Totem groups claimed a similar freedom of expression in furniture, creating bright, convoluted tables and chairs whose fabrication combined both manual work and new technology. Galleries showed them as artworks. Moreover, the growing emergence

The boundary between art and design is sometimes blurred. This is the subject of Robert Stadler's work from 2004 (Pools 8 Pouf!, courtesy Galerie Dominique Fiat).

of installation works as a new form of artistic expression asked the viewer to look at everyday objects differently: artists used them for their form, color, and material rather than their function. Tony Cragg, Sarah Sze, and Gilles Barbier created strange and fascinating works like this. Artists may even devise furniture: Bernard Rancillac's elephant armchair and Hubert Le Gall's flowerpot chairs are memorable examples.

Over the last few years, the relationship between art and design has given rise to a questioning of each discipline's practices and vocabulary. Sometimes design objects are presented next to artworks in exhibitions.

This is the case, of the Lit Clos (Closed Bed) by the Bouroullec brothers, the Pools & Poof poufs by Robert Stadler, which are in turn couches, rugs, or wall decorations, and the Whippet bench in the form of a whippet dog by the Radi Designers. As for artists, they readily play with design models, appropriating design's visual impact and giving another interpretation of designer classics. Franz West took the first step: his couchsculptures covered by kilim rugs can also be used as chairs. Kimiko Yoshida's decorated mirror tables are both paintings and sideboards. By turning Marc Newson's famous Embryo armchair upside down, Bertrand Lavier makes it into a bull's head in homage to Picasso, while Richard Fauguet reinvents Marcel Breuer's famous tubular armchair by over-enlarging it and recreating it out of drainpipes. By way of a signature, Philippe Cazal commissions a graphic design agency to make a logo out of his name, and Franck Scurti reinterprets road signs. Within the context of the relational aesthetics movement, which introduces an interactive relationship between the spectator and the artwork, Rikrit Tiravanija creates collective, convivial, and functional spaces, situated at the crossroads between art and design, which have decidedly never been so close.

By turning designer Marc Newson's famous armchair upside down, the artist Bertrand Lavier repeats the gesture of Picasso, which transformed a set of bicycle handlebars and seat into a bull's head (limited edition at Galerie Kreo, 2002).

FACING PAGE
For the artist Olivier Liégent, painting can adapt to all uses (Living Paintings, 1999).

DESIGN AND PUBLIC SPACES:
beautiful adornments

In the 1960s, when the world was separated into two blocs, one communist and the other capitalist, involved in a bitter ideological war, the Western tourist visiting Moscow for the first time felt as if he had arrived on a different planet. The total absence of advertising posters, the rare products clumsily presented in the windows of state-run stores, and the handful of antiquated cars driving around made him feel as if he had gone back in time. Novels set in early twentieth-century Paris describe with pleasure the metro entrances designed by Hector Guimard in pure art deco style in 1900 or the famous advertising pillars of the Grands Boulevards. What would London be like in the collective imagination today without its double-decker buses and red telephone booths, or Amsterdam without its green streetcars and black bicycles? It's a fact: the elements that make up the urban landscape play a role in the postcard impression we have of a city.

From the mid-nineteenth century, newspaper kiosks, lampposts, and street name plates began to fill the modern city with their signs.

Today, billboards, road signs, signposts, bus stops, pay-and-display machines, and trash cans impose a hi-tech homogeneity everywhere we look, dictated by the reign of the automobile. However, since the 1980s, change has been afoot. Planning offices staffed with designers, engineers, and technicians propose alternatives to this all-purpose style, in close collaboration with the public authorities. For example the JCDecaux company regularly calls upon globally renowned designers to design better adapted urban furnishings. When designed by Sir Norman Foster, the bus shelter manages to blend elegantly into both modern and old cities. Moreover, the style of cafés, restaurants, hotels, and meeting places all testify to the spirit of the times: The Florian in Venice, the Deux Magots in Paris, the Einstein in Berlin, and the Sprüngli in Vienna all now belong to history.

A new phenomenon, one that first saw the light in the 1980s, is spreading like wildfire: designers all over the world are being commissioned to decorate places that may one day become iconic.

The bar at the Hôtel Plaza Athénée in Paris, designed by Patrick Jouin in 2001, exudes a dreamlike and theatrical atmosphere.

Philippe Starck, one of the designers most active in this area, designed the Café Costes in Paris, the Manin in Tokyo, and the Royalton Hotel in New York, to name but a few. Andrée Putman made a big hit with the refurbishing of the Morgans Hotel, also in New York. In Barcelona there is Alfredo Vidal Alvarez's KGB and in Madrid the Hotel Puerta América where each floor is designed by a designer of international fame. Both reflect the country's economic expansion and chime with the Movida spirit. The sci-fi decoration of the Georges restaurant situated on the top floor of the Centre Pompidou in Paris, designed by Dominique Jakob and Brendan MacFarlane, has been much discussed over the last few years. As has the Hi Hotel in Nice, designed down to the smallest teaspoon by Matali Crasset.

Commissioned by JCDecaux in 1994, the elegant lines of Norman Foster's bus shelter means it melts into the urban landscape.

These initiatives have increased the popularity of establishments who have thrown themselves into the adventure. When beauty rhymes with business.

Andrée Putman's
sober and elegant
refurbishment
of the sixteenth-
century Lainé
warehouses
in Bordeaux has
contributed
to the fame
of the city's
CAPC museum
of contemporary
art (1990).

FACING PAGE
Designed by
Inga Maurer,
the entrance
to this hotel
in Maastricht
creates an ingenious
bridge between
past and future
(Kruisherenhotel,
2003).

IT'S A GENERATION THING!

IT'S A GENERATION THING!

IT'S A GENERATION THING!

DESIGN IN THE 1920s
a change of style

Red and Blue, Gerrit
T. Rietveld, Cassina
Maestri collection,
1918.

FOLLOWING PAGES
At the Musée des
Arts Décoratifs in
Paris, a faithful
reproduction of the
boudoir adjoining
the fashion designer
Jeanne Lanvin's
bedroom designed
by Armand-Albert
Rateau in 1924-25.

the era

At the end of the First World War, people wanted to forget the horror of the fighting and live it up a little. In the Roaring Twenties, everything was an excuse to party. Women, who had taken on new responsibilities while the men were at the front, now claimed a new status and asserted themselves with a more provocative look: short skirts, fluid lines, and bobbed hair. Fashion designers were quick to translate the spirit of this era in which radios, automobiles, and trains contributed to the acceleration of a change in lifestyle.

the style

A style of simple, geometrical lines emerged in response to art nouveau, which was judged too baroque. The furnishing industry developed: the French department stores Galeries Lafayette and Bon Marché sold items specifically designed with mass production in mind. Art reviews and decoration magazines published articles on interior design with designer furniture made in quality materials such as ebony covered in parchment or lemonwood inlaid with brass threads and Brazilian rosewood. In France, Ruhlmann's furniture represented the quintessence of luxury, while Jourdain and Chareau made furniture inspired by modernist architecture. Encouraged by the notion of Communist utopia and fascinated by industrial production, the Bauhaus school in Germany was a hotbed of experimentation during the same period, turning out furniture prototypes using innovative forms and materials.

key pieces

◦ The cozy corner
◦ Marcel Breuer's metal armchair, 1925
◦ Eileen Gray's Satellite Mirror, 1927
◦ Pierre Chareau's Brazilian rosewood chaise longue, c. 1923
◦ Louis Sognot and Charlotte Alix's bar stool

recurrent themes

◦ Inlaid lozenge-shape friezes
◦ Ornaments: flowery bowls, medallions with flowers or geometrical motifs
◦ Tapered or lotus legs, with cut or fluted panels
◦ Leather, parchment, or shagreen leather casing

As the French song goes "It was the time of the Roaring Twenties, of limousines and the Charleston." After the war, there was an urgent desire to move on to other things. The style of the 1920s breaks with the past: geometrical forms, smooth and shiny materials, and discreet ornaments symbolizing the modern spirit.

1 Closet in Macassar ebony, tortoiseshell, and ivory, Jacques-Émile Ruhlmann, c. 1920.
2 Black lacquered, glass, and silvered bronze bureau bookcase, Jacques-Émile Ruhlmann, 1929.
3 Brazilian rosewood day bed, Pierre Chareau, c. 1923.

4 Silver cabinet known as the Meuble au Char (Chariot Furniture) Macassar ebony and ivory, Jacques-Émile Ruhlmann, 1922.
5 Satellite mirror, Eileen Gray, 1927.

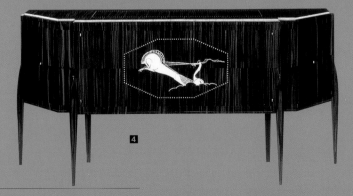

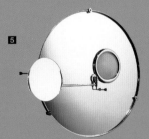

DESIGN IN THE 1930s
the triumph of art deco

Telephone (LMT), 1930.

FOLLOWING PAGES
The Compagnie des Arts Français created this bathroom in 1938 at the French Ministry of Foreign Affairs on the occasion of the visit of King George VI of England and his wife Elizabeth to Paris.

the era

Modernity entered the apartment thanks to the widespread use of electricity. Recent buildings, built in concrete, had large bay windows which opened up to the outside world. The living room became a place in which to live and receive visitors. The clean lines of art deco furniture were a must, all the rage with the happy few able to travel to the United States on the splendid ocean liners linking Europe to the New World, which offered pure art deco decoration signed by the likes of Ruhlmann, Dunand, Leleu, Sue, and Mare.

the style

The art deco style is characterized by cubist geometrics and the architectural aspect of its furniture design. The luxurious finishing touches and the quality of the materials contrast with extremely discreet patterns and ornamentation. This comfortable, refined, and elegant furniture is at home in a smooth-walled, bright environment. Avant-garde designers continued their research and prototypes that had appeared a few years before, in particular metal and leather furniture pieces, began to be produced in small series. Le Corbusier, who founded the Union des Artistes Modernes (UAM) in Paris for the theorization of this radically new practice in 1929, grew in fame and influence.

key pieces
❖ Jean-Michel Frank's adjustable shelves
❖ The bar
❖ The lounger by René Herbst
❖ The magazine rack
❖ The reclining chaise longue
❖ The low couch

recurrent themes
❖ Lacquered wood
❖ The search for precious woods and the highlighting of their natural graining
❖ A return to ebony
❖ The use of metal and glass

In the 1930s, innovators pursued the utopian notion of beautiful furniture accessible to all. They considered furniture to be a continuation of architecture. Although they succeeded in manufacturing the fruits of their research in small series, their creations and in particular those in metal, judged to be too austere, found little support among the general public.

1 Director desk, Jean-Michel Frank, 1930.
2 Beolit 39, the first in a series of radios molded in Bakelite, Bang 8 Olufsen Team, 1938.

3 Auto-Thermos pressure cooker, around 1930, Boulogne-sur-Seine workshops.
4 Salon d'Automne, Paris, Le Corbusier, Pierre Jeanneret, and Charlotte Perriand, 1929.

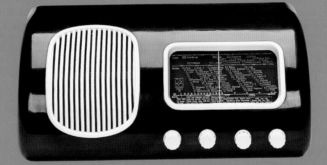

DESIGN IN THE 1940s
traditional versus modern

*Laurent telephone,
plastic, 1941.*

the era

If the war brought the furniture industry to a halt in France, it pushed other countries to invest in technological research. Bauhaus migrants who arrived in the United States in the 1930s had already started working with new materials such as bent plywood, molded plastic, and rubber. When the fighting came to an end in Europe, reconstruction became an urgent issue. The creation of residential estates with the massive production of cheap apartments overturned the conception of living spaces. The term "design" was not yet in common use and the situation seemed somewhat confused: on the one hand there were those nostalgic for the styles of the past and the unique piece of work, and, on the other, there were the moderns, sensitive to social problems and the democratization of things that make up everyday life.

the style

The 1937 Exposition Universelle des Arts et Techniques in Paris and the *Tomorrow's World* exhibition in New York in 1939 sowed the seeds of a resistance that would persist in France over the next ten years. The old generation, supporters of the "belle tradition française" (fine French tradition) and artistic cabinet work, began to design sophisticated furniture which both referred to and simplified the Louis XV, Louis XVI, and Restoration styles, for the decoration of the huge rooms of national palaces. Then there were the innovative creators—Prouvé, Royère, Sognot—who worked closely with architecture. They addressed a less wealthy clientele, and defended functional furniture with simplified lines and no decoration. Natural wood, plywood, metal, rattan, and linoleum were the favorite furniture materials of the Reconstruction era. In the United States, Charles Eames and Eero Saarinen started inventing tomorrow's furniture.

key pieces

❁ The cane shell chair
❁ The lacquered chest of drawers
❁ The plywood chair
❁ The Polar Bear armchair and couch by Jean Royère, 1946
❁ Jean Prouvé's low pedestal table, 1944
❁ Raymond Loewy's redesign of the Lucky Strike cigarette package, 1940
❁ Henry Dreyfuss's Hoover vacuum cleaner, 1949
❁ Tetra Brick packaging, Tetra Classic, 1944
❁ Raymond Loewy's Coca-Cola drinks vendor, 1943

recurrent themes

❁ Plywood
❁ Wooden bases
❁ The veneering of precious woods (sycamore, mahogany, ebony, Brazilian rosewood)
❁ Inlaid copper, mother-of-pearl, and ebony
❁ Bronze brackets and caster sockets

War once again raged across Europe. Materials were rare during this period of withdrawal and restriction. But in the United States, where the most radical designers were received with open arms, innovation accelerated. However, everyone had to choose his camp: on both sides of the Atlantic discord between the traditionalists and the modernists was on the rise. And the 1940s style contained the seeds of this rift.

1 Arpeggio radio, Bakelite, wood, glass, and fabric, Philips, 1942.

2 Molded Bakelite office intercom, Livio and Pier Giacomo Castiglioni, 1939–40.

3 Turnover toaster, Kenwood Manufacturing Company, 1947.

4 Polar Bear couch, Jean Royère, 1946.

1

2

3

4

DESIGN IN THE 1950s
domestic science liberates women!

Diamond chair,
Harry Bertoia,
Knoll, 1952.

the era

With the baby boom and the return of prosperity, the post-war period marked the beginning of consumer society. After the invention of nylon stockings, first worn in the United States, women now benefited from the arrival of electrical appliances. Which, as the commercial says, liberate women! Radio and television brought news into the home of a world regularly threatened by the rising tensions of the cold war. The dining room became a "living room," a symbol of the family's togetherness and shared living space. Designers explored a new "art of living" tinged with humanism and conviviality. Across the Atlantic, the manufacturing of pieces in small series by Herman Miller and Knoll contributed to design's popular recognition. The Ulm school in Germany, founded in 1955, attempted to continue the teachings of the Bauhaus.

the style

Many new designers who were eager to exploit the novelty and innovation available to them after the lean war years found opportunity and inspiration at the Festival of Britain in 1951. Scandinavian designers like Finn Juhl and Eero Saarinen were designing practical pieces privileging rounded forms and natural materials. Meanwhile in the United States, husband and wife duo Charles and Ray Eames were the first to work with bent plywood and plastic, ushering in a new era of affordable, mass-produced furniture featuring bright colors and soft, modern shapes. Space gain was a priority at this time and so furniture was rationalized. The 1950s style featured sober, Spartan lines and smooth surfaces with orthogonal or less rigid forms on metallic legs. It was the reign of the Formica table. New materials and the development of manufacturing techniques such as molding gave rise to adjustable, functional furniture. Partitions slid, chairs stacked, and furniture of various uses lined up along the walls.

key pieces

✿ Jean-Prouvé's desk-table
✿ Mathieu Matégot's bunk bed
✿ Serge Mouille's three-armed floor lamp
✿ The storage unit
✿ The extension-leaf dining table
✿ Arne Jacobsen's Ant chair, 1952
✿ Harry Bertoia's Diamond armchair, 1952
✿ Kenwood mixer, eight accessories, 400 kW power, 1950
✿ Isamu Noguchi's Akari 10A lamp, 1951
✿ Charlotte Perriand's Shadow chair, 1955
✿ Braun's SK5 radio transistor and record turntable, 1956
✿ The Moulinex coffee grinder, 1959

recurrent themes

✿ Plastic or molded wooden laminated monoshell structures
✿ Tubular steel legs
✿ Formica and laminated materials
✿ Plastics

The 1950s seemed to emerge from a long period of hibernation. And the housing problem urgently needed a solution. Functional furniture inspired by industrial furnishing grew in popularity. New materials and innovative processes allowed for the creation of new shapes adapted to new living spaces. Electrical appliances provoked changes in domestic behavior.

1 Three-branched floor lamp, Serge Mouille.

2 The first Chef A700, Kenwood Manufacturing Compagny, 1950.

3 The first Moulinex coffee grinder, 1956.

4 SM3 electric razor, Braun, c. 1950.

5 Compass desk, Jean Prouvé, 1950.

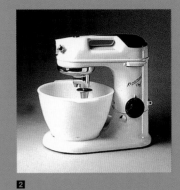

LE XXI° SALON DES ARTS MÉNAGERS

Si depuis sa fondation le Salon des Arts ménagers a connu un succès toujours croissant, c'est qu'il répond assurément au goût de notre époque éprise de confort, de simplification. Les Français gardent l'amour du foyer dont l'aménagement permet à la personnalité de chacun de s'affirmer. Ce foyer, ils le veulent commode, équipé de manière à épargner à la maîtresse de maison les travaux pénibles. Pour eux le Salon des Arts ménagers est un réservoir d'idées ingénieuses et de solutions pratiques. Et le XXI° (voir notre article), comme ses devanciers, emplit le Grand Palais.

FACING PAGE

The Novo canteen with Ant chairs, by Arne Jacobsen, Fritz Hansen, 1952.

France Illustration magazine cover, March 15, 1952.

1960s advertisement with Egg seats by Arne Jacobsen, Fritz Hansen.

IT'S A GENERATION THING!

DESIGN IN THE 1960s
burning with enthusiasm

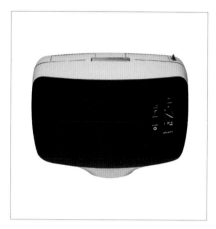

*Portavia P111
television, Roger
Tallon, 1964.*

the era

The early 1960s were euphoric: the economy was growing and full employment ensured an optimistic vision of the future. A new art of living, based on the democratization of consumer goods, changed the relationship with the everyday environment. Increasingly, popular interior decoration magazines provided an echo of this more playful relationship with domestic space. Parisian department stores increased their furniture shows. Some even offered their own catalog (Prisunic was the first to do this). The break with functional forms preceded the arrival of Pop culture. Born in 1966 in the United States, the hippie movement spread its thirst for social freedom. The number of demonstrations in favor of utopian ideals and against consumer society grew steadily before reaching its climax in France in May 1968. While at Woodstock or on the Isle of Wight, the young dreamed of tomorrows full of song and communal living against a background of hallucinogenic drugs, everything was also allowed in design.

the style

During this period of unbridled creation, plastic quickly became an ideal material for designers. Its malleability meant it could be experimented with for even the craziest of desires. The metallic structure of chairs disappeared under a thick foam layering. Sometimes, as in the case of the futuristic inflatable armchair, it simply does not exist. Curves and colors ruled. A famous example of this is the Djinn series by Olivier Mourgue, chosen by Stanley Kubrick in 1968 to furnish the film set of *2001: A Space Odyssey*. Humor and derision were now an integral part of furniture intended for communal living and adapted to the body. People relaxed close to the ground, curled up in attractive couches, or even in armchairs cast from the human body. In the middle of all this agitation, French design attained recognition during the Fourteenth Milan Triennale.

key pieces

- The Globe armchair by Eero Saarinen, 1968
- The Culbuto armchair by Marc Held, 1967
- The Tam Tam stool by Henry Massonnet, 1968
- Inflatable couches and armchairs by Quasar Khanh, 1968
- The film set of *2001: A Space Odyssey*, furniture by Olivier Mourgue, 1968
- The Marie food processor by Moulinex, 1961
- The Sixtant SM31 electric razor by Hans Gugelot and Müller, Braun, 1962
- The Portavia P111 television by Roger Tallon, 1963
- The portable transistor radio by Bang & Olufsen, 1965
- The Instamatic by Kenneth Grange, Kodak, 1966

recurrent themes

- The curved line
- Soft forms
- Corrugated foam
- Latex
- Elastic jersey
- Bright colors

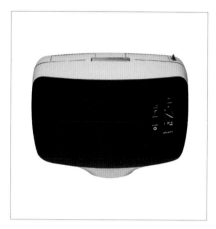

IT'S A GENERATION THING!

The baby boomers were now twenty years old, the age when every wild and crazy thing is possible. Consumerism continued to soar, scarcely touched by the debates it provoked. Life was great. An air of joy and freedom was blowing in all fields of creation, and the plastic fetish became associated with the colorful and zany style of the 1960s.

1 After the Dauphine automobile, Renault launched the 4L.

2 Culbuto armchair shell in reinforced polyester and molded fiberglass, lined with foam and covered in fabric, Marc Held, Knoll International, designed in 1967, 1970.

3 1969 advertisement for the first Moulinex mixer (1959).

4 Diabolo or Tam Tam stool, Henry Massonet, 1969.

5 2001: A Space Odyssey, Stanley Kubrick: the space station's Hilton room, 1968.

1

2

3

4

5

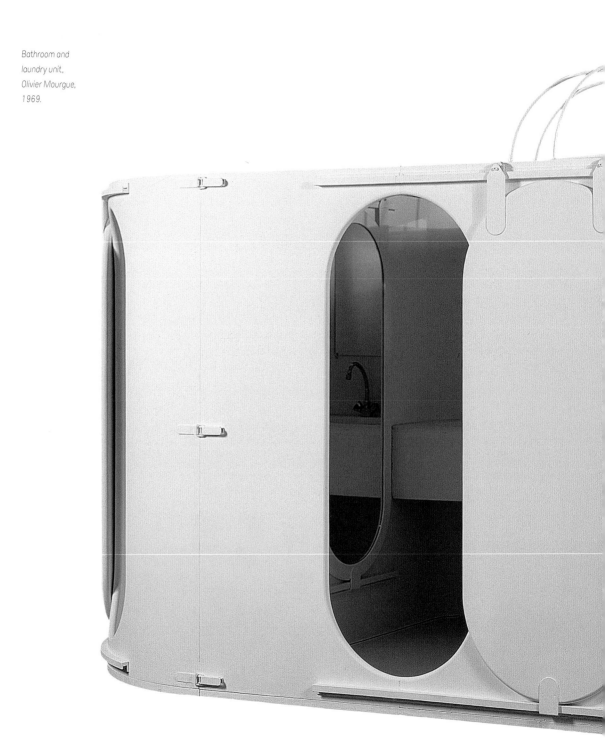

Bathroom and
laundry unit,
Olivier Mourgue,
1969.

DESIGN IN THE 1970s
back to sobriety

Terraillon kitchen
scales, Marco
Zanuso, 1970.

the era

The breath of freedom was felt for a while after 1968, but was cut short by two oil crises which soon cast a shadow over the beatitude of a spoiled generation. Global economies were shaken and protest movements stifled. Very quickly, France joined the technological race. Between 1972 and 1979 the prototype of the TGV high-speed train, Roissy Airport in Paris, the microchip card, Concorde, and the Ariane rocket were developed in quick succession, with the symbolic Centre Pompidou built by Renzo Piano, Richard Rogers, and Gianfranco Franchini in the heart of Paris. In the United States, New York artists rehabilitated large-scale industrial spaces and transformed them into lofts in an assertion of a hi-tech style in stark contrast with the Pop years. However, Milan proved to be the world capital of design thanks to the commitment of dynamic Italian manufacturers who supported a new generation of talents ready to fight against modernism's rigid aesthetics.

the style

The pessimism of the 1970s manifested itself in two distinct design perspectives: Anti-Design and the High-Tech movement. Proponents of Anti-Design, most apparent in Italy, produced intentionally awkward or kitsch pieces featuring bright, bold colors in defiant revolt against the austerity of Modernism. Meanwhile, American followers of High-Tech urged a return to the timeless designs of Modernism which they claimed would reduce manufacturing costs and curb the environmental damage brought on by the excesses of consumerist society. Furniture became a product for the general public, made in a neutral style: furniture kits to take away with tubular structures. It was the era of large-scale distribution and mail ordering. After 1976, with the rapid rise in oil prices and the emergence of new ecological movements, there was a return to the safe material of wood. The rustic country style and its solid pine shelves met with public approval.

key pieces

* The Poltrona di Proust armchair by Alessando Mendini, 1978
* Pierre Paulin's decoration of the Elysée Palace offices, 1972
* Furniture kits
* Modular furniture
* The 3 Suisses catalog
* Habitat, Ikea
* The Terraillon kitchen scales by Marco Zanuso, 1970
* The espresso coffee machine by Alessi, 1979
* The Togo model by Michel Ducaroy, Ligne Roset, 1973
* The first Walkman, model TPS-L2, Sony, 1979

recurrent themes

* Furniture cut in blocks
* Pine
* Tubes and canvas
* Altuglas
* Colors: orange and brown

IT'S A GENERATION THING!

The party spirit and the search for alternative lifestyles persisted into the middle of the decade. But the oil crisis was to bring utopian notions to a grinding halt. The designers of the 1970s shift between vaguely futuristic research and the designing of practical and sober furniture sold for the first time in kit form at affordable prices thanks to large-scale distribution.

1 Valentine typewriter, Ettore Sottsass and Perry A. King, Olivetti, 1969.
2 The first Sony Walkman, TPS-L2, 1979.
3 Pratone, rug-seat, cold polyurethane foam, Guflex in the shape of a giant lawn, and Guflac washable paint, Gufram, 1971.
4 Ikea catalog, Sweden, 1978.
5 Togo range of seats, foam, polyester fiber, and quilted Oxford fabric, Michel Ducaroy, Ligne Roset, Gruppo Strum, designed in 1973, 1976.

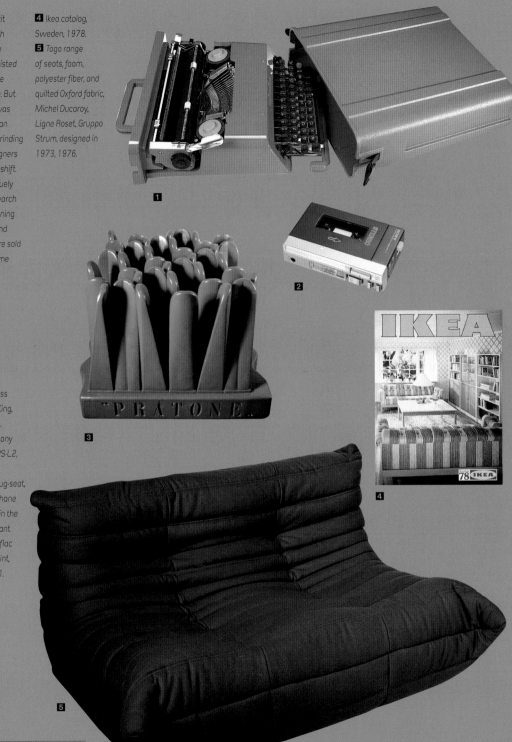

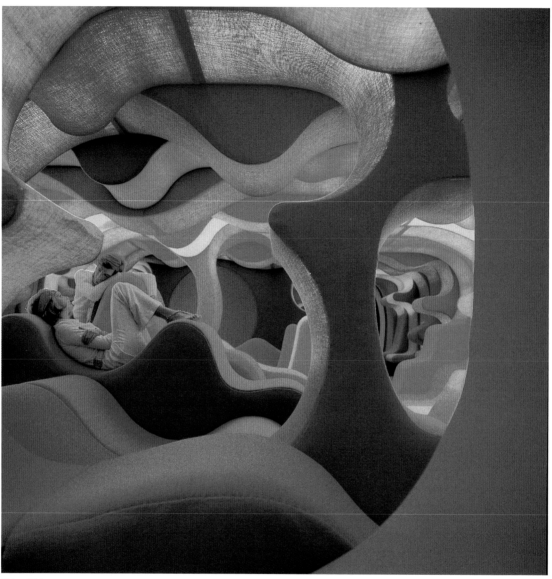

Visiona 2 living room,
Verner Panton,
Cologne furniture
trade show, 1970.

FACING PAGE
Élysée Palace
Smoking Room,
Pierre Paulin, 1972.

DESIGN IN THE 1980s
unbridled creativity

Super lamp,
Martine Bedin,
Memphis, Milan.

the era

With the discovery of the HIV virus, the fall of the Berlin Wall, and the Chernobyl catastrophe, the 1980s were a period of agitated transition. Everything was turned upside down. A new economic dynamism emerged from the effects of globalization and the Internet. Personal success was flaunted, while the widespread use of personal computers revolutionized the concept of work. "Radical design" or "new design" by Italian groups such as Studio Alchimia and Memphis, who advocated decoration for decoration's sake and a return to spontaneity, were all the rage. The postmodernist era began under the sign of a creative experimentation without hang-ups.

the style

Bubbling, eclectic, and contradictory, the style of the 1980s refuses to be contained within fixed definitions. This is exactly what its protagonists wished for. The future already seemed out of date. References, periods, and materials overlapped to create concept furniture. Objects were not meant to be comfortable or functional, but to announce (or denounce) a discourse, formulate an inquiry, or poke fun. If sharp lines and bright colors abounded, it was because a new formal world, full of fantasy, arose from this continual questioning, the result of mixed symbols and geographical, geological, or historical deviations. In France, the Garouste and Bonetti duo was a good illustration of this. However, this effervescence also favored the emergence of figures such as Starck, Dubuisson, Szekely, and Wilmotte who were more interested in the simplified lines of the working drawing.

key pieces
- The leather couch
- The Palais Royal armchair by Wilmotte, 1986
- The metal sheet and stone triangular table by Garouste and Bonetti, 1983
- The Zebra armchair in lacquered wood by Totem, 1982
- The Miss Blanche armchair by Shiro Kuramata, 1989
- The Carlton shelves by Ettore Sottsass, 1981
- The Swatch
- The Macintosh SE by Frog Design and Apple, 1987
- The Super lamp by Martine Bedin, 1981

recurrent themes
- Zigzag lines
- Flecked or printed coverings
- Mirror splinters, object fragments
- Lacquered wood

Economic globalization and the revolutionary appearance of the Internet created a new turning point. Designers freed themselves from the dictates of functionalism, influential since the 1920s, and dipped into various geographical and historical references in order to create an eclectic and baroque style.

3 9093 kettle, Michael Groves, Alessi, 1985.
4 1989 desk, Sylvain Dubuisson, purchased by the Ministry of Culture, Fourniture, 1989.
5 Feltri armchair, Gaetano Pesce, Cassina, 1986.
6 Palais Royal chair, Jean-Michel Wilmotte, Academy, 1986.

1 First Swatch collection launched in 1983.
2 Jupiter Totem stool, lacquered metal and turned, lacquered wood, Jacques Bonnot, 1982.

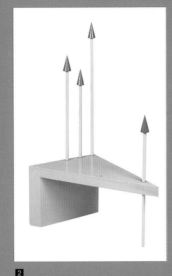

1

2

3

4

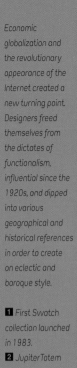

5

6

FACING PAGE
Café Costes,
Philippe Starck,
Paris, 1984.

Bathroom,
Andrée Putman,
Morgans Hotel,
New York, 1984.

DESIGN IN THE 1990s
a soft touch

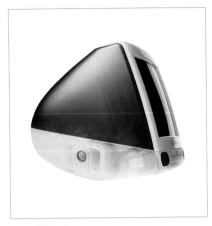

iMac 1998, Apple.

FOLLOWING PAGES
Georges restaurant on the sixth floor of the Centre Pompidou, Jakob and MacFarlane competition project in 1988, built in 2000.

the era

The decade began with a new kind of conflict. The Gulf War, its precision air strikes filmed and broadcast almost in real time, resembled a video game. The explosion of the Internet and the "miraculous" new economy were not far off. However, in the light of a far from reassuring future, design was tinged with a certain humility. This may seem paradoxical, at the very moment when the media coverage of designers took off. Hotels, restaurants, boutiques: everyone wanted to work with them. Henceforth, they were wooed by important industrial firms anxious both to stand out and give character to their products. And that's not all, for didn't Philippe Starck intend to create "objects that like people?"

the style

Exuberance is out. Clean lines are in, and correspond to a desire for an economy of means. Furniture calls for simplicity, mixing organic with geometrical forms. A new vocabulary dominated by the curve and the contour is developed. Whether it concerns the design of a train, an automobile, or a household utensil, the object addresses the five senses. Better yet: not only is it interactive, but also "intelligent," thanks to its design by computer and the development of composite materials. Another priority is that of preserving the environment, either by prolonging the life of the components, or by making them 100 percent recyclable.

key pieces
- The Nomos table, Norman Foster, 1989
- The Honda vase, Ronan and Erwan Bouroullec, 2001
- The Embryo chair, Marc Newson, 1988
- The Juicy Salif lemon squeezer, Philippe Starck, 1990
- The Mayday hanging lamp, Konstantin Grcic, 1998
- The Sim chair, Jasper Morrison, 1999

recurrent themes
- The return of the curve
- Translucency
- Acid colors
- Molded polypropylene
- Composite or "intelligent" materials

IT'S A GENERATION THING!

Exuberance was coming to an end: it was now a time of gentle lines and simple shapes. But also of ecological awareness: the earth's riches are not inexhaustible. Recycling and salvage became popular with designers inspired by the practice of artists. Furthermore, the iPod and other technological materials change our way of life.

1 April Vase, glass and galvanized steel, Tsé & Tsé, 1991.

2 BeoSound 9000, audion CD player, a visual concept with a set of 6 CDs, David Lewis, Bang & Olufsen, 1996.

3 Mayday lamp, polypropylene, Konstantin Grcic, Flos, 1998.

4 Orgone II, blue rotomolded plastic monobloc, Marc Newson, Marc Newson, London, United Kingdom, 1998.

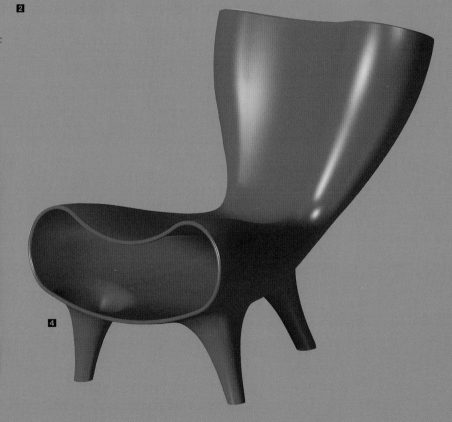

1

2

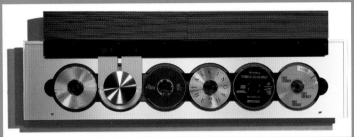

3

4

DESIGN IN THE YEARS 2000
ostentation meets sobriety

*Vegetable Matrix,
woven living black
willow, Godefroy
de Virieu (VIA).*

FOLLOWING PAGES
*Mama Shelter hotel,
artistic director:
Philippe Starck,
architect: Roland
Castro, 2008.*

the era

Everything begins and ends with Wall Street. The dawn of the year 2000 excited the imagination. The new millennium offered a thousand promises: the global economy must guarantee long lasting peace and a fairer balance.

But September 11, 2001, cooled this utopian outburst: the collapse of the Twin Towers in New York, shown on loop, almost in real time, on all the televisions of the world, presented a surrealist spectacle with tragic human and political consequences. America called for revenge, tried to curb the Middle East, invaded Iraq, and sank into a long war. At the same time, ecologists sounded the alarm while the stock exchange underwent a rapid and almost indecent rise. Designer goods made record benefits. The "bling-bling" aesthetic took over: nothing was too beautiful, too flashy, or too expensive. At the beginning of 2008, a terrible stock market crash forecast the end of such prosperity. The fall of the financial markets led to a serious economic crisis.

the style

Eclecticism reigns supreme. The designer becomes also a decorator. He combines tendencies, concocts interiors with a contemporary feeling, and dramatizes public spaces. Sobriety rubs shoulders with exuberance. It is a time of mixings: the moderns are updated, styles and periods are brought closer, and tradition is revisited and revised. Ethnic objects are rehabilitated, hybrid materials are created, origins are mixed. Simplicity on the one hand, ostentation on the other.

IT'S A GENERATION THING!

key pieces

- The Cristal Room, Philip Starck, Baccarat, Paris
- Legend bookcase, Christophe Delcourt, Roche-Bobois
- ChairOne-4 Stars, Konstantin Gric, Magis
- Outdoor pouf, Fatboy
- Cork Family stool, cork, Jasper Morrison
- Salad bowls, Ekobo
- Fluorescent clogs, Crocs

recurrent themes

- Articulated, modular couches
- Weaving
- Origami
- Flat screens
- Ultra-miniaturization
- Eco-design

Caught between ecological awareness and the desire for flashy works, designers became schizophrenic: projecting into the future to play with high technology and its innovative processes, while dreaming of elementary objects with sober, Japanese forms in natural and recycled materials. Bamboo, concrete, felt, and wood as well as Corian, crystal, and gilded metal were very popular. Designers turned into artists and vice versa. The designer of the year 2000 was looking for new ethics. High technology meant he could solve problems of the recycling and wastage of natural materials. He became an artist and invented constantly evolving forms. Objects and furniture must respond to the changing mood of the time.

1 Spaghetti bench, Pablo Reinoso.

2 Adjustable and articulated Flap bed, Francesco Binfaré, Edra.

3 Nano iPod, Apple.

4 Gold bar, Arik Levy, Eno.

5 Bedside lamp from the Gun collection, Philippe Starck, Flos.

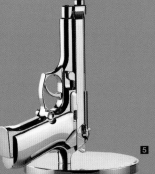

WHAT'S YOUR STYLE?

WHAT'S YOUR STYLE?

are you HI-TECH?

You hate flowery cushions, thick pile carpets, swan-neck taps. You believe that wall moldings and ceiling paintings belong to another era. You prefer open spaces to closed places, bright light to dimmed light, practical elements to superfluous details. You dream of a generous society without hierarchy or protocol. And you like simple lines, functional forms, and easy-to-clean materials. Your non-conformism therefore naturally pushes you toward hi-tech style.

What is particular about it? A capacity to convert, in the home environment, equipment, or industrial materials originally designed for communities or professional spaces such as schools, hospitals, offices, and army barracks. This means it is foolproof. More than just a fashion, hi-tech is one of the more important movements in contemporary design.

A contraction of the terms high style and technology, it appeared on the interior decoration scene in the late 1970s, in the wake of the fashion for lofts and living spaces constructed in abandoned workshops, factories, and warehouses. Contrary to what the term implies, the movement does not concern the latest technological must-have when referring to interior design. Let's just say that a faint whiff of nostalgia floats over even the most sober hi-tech interior designs. For the desire to bring furniture and objects linked to the workplace into the home does in fact reactivate early twentieth-century functional aesthetics conceived by designers concerned with meeting the requirements of economy and safety. Moreover, hi-tech refers to the heroic times of the first prefabricated buildings. The purists fix the date of its birth in 1851, when Sir Joseph Paxton built the huge Crystal Palace in the space of a few months, which then served as the entrance hall for the international exhibition in London. This gardener-engineer had the idea of using the system for building greenhouses, through a quick method of assembling prefabricated and pre-cut castings, wrought iron, and glass. An architectural gesture that the innovative-minded greeted as one of the most beautiful demonstrations of the potential dialogue between art and technique. Furthermore, in the 1920s, the international modern movement will see in this style, despite its lack of specific aesthetic research, a complete and violent break with lower middle-class tastes. The architect Le Corbusier, who saw the house as a fully equipped "machine for living in," was one of its most enthusiastic supporters. As for Communist revolutionary ideology, it was dreaming then of a classless domestic environment.

Later, in New York in the 1950s, impecunious artists and intellectuals reconverted equipment from the warehouses and workshops where they were squatting. The enameled lamp shade in the form of an upside-down dish, hospital faucets, laboratory trolleys, factory neon lights, metal staircases, and armor-plated doors have since become a must with the new bourgeoisie who prefer flicking through a DIY catalog to the traditional glossy interior design magazine.

Today, with the miniaturization of electronic components, the term "hi-tech" is used to describe the latest digital gadgets. However, materials and equipment from the first industrial age have now become icons which inspire designers.

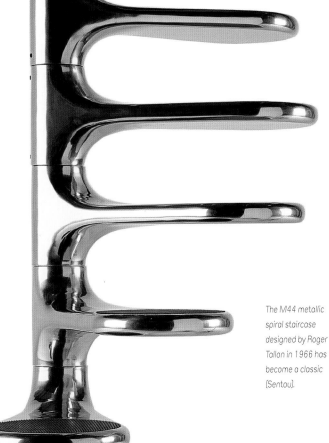

The M44 metallic spiral staircase designed by Roger Tallon in 1966 has become a classic (Sentou).

Roger Tallon, for example, devised a spiral staircase which is now considered a classic of the metallic aesthetic. And Shiro Kuramata amused himself with How High the Moon, which revisits the club chair in a wire-netting version. As for Marc Newson's famous polished and gleaming chair, it resembles an elegant sheet of curved steel. In more prosaic terms, Rody Graumans' 85 Lamps ceiling light for Droog Design brings together 85 light-bulbs just as they are. It is difficult to be more basic and less sophisticated.

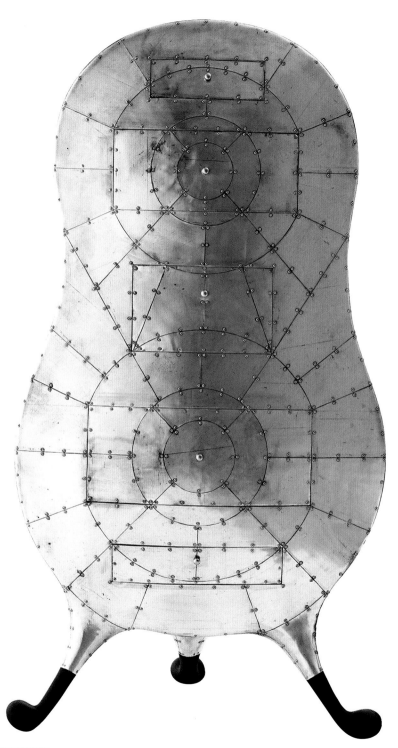

FACING PAGE
This club chair, made out of ultra-light and hard-wearing wire netting, was designed by Shiro Kuramata (How High the Moon, 1987, Vitra).

A combination of round, sensual forms and allusions to science fiction, this wardrobe by Marc Newson evokes the adventures of Captain Nemo (Pod of Drawers, 1987, Pod).

are you BAROQUE?

Functional furniture? No, thank you. There is no way you'll live in a kitchen! You reject incantatory diktats, rigid dogmas, and fundamentalist theories. You militate in favor of free expression and individual happiness. The cold, uncompromising rigor of minimalist interiors distresses you. You like to saunter through interiors full of precious trinkets, overly ornate furniture, shimmering fabrics, and rare objects. Faced with the challenges of the twentieth century, you feel it is urgent to give the imagination free rein and to transform the house into a refuge full of fantasy. In short, confronted by the trivialized, soulless forms flooding department stores, you defend extravagance and creative expression. The neo-baroque movement should suit you to a tee.

At the turn of the 1980s, the return to symbols, hedonism, and ornamentation in design was a bombshell. If not a total betrayal of the sacred modernist concepts erected in the 1920s. It all started in Milan in 1979, at the furniture show: the public was immediately charmed. A team of young, impetuous designers grouped around Alessandro Guerriero and including Ettore Sottsass Jr., Andrea Branzi, Alessandro Mendini, Michele De Lucchi, and Paola Navone, united under the name of Studio Alchimia, proposed deconstructed cupboards, tables, and chairs straight out of a comic strip. This manifesto-furniture by the Italian rebels marked the passage from modernism to postmodernism and was shown in art galleries. Studio Alchimia, then Memphis, opened up a gap to be filled by countless designers. In France, at the same time, while the Totem group were brightening up the domestic environment by revising historical and exotic forms, an avant-garde movement called Les Nouveaux Barbares appeared. Swept along by Élisabeth Garouste and Mattia Bonetti, this movement reinvented, in its own way, the tradition of Parisian upholsterer-decorators. The famous duo created iconoclastic furniture that mixed all sorts of materials: pebbles, stretched hides, beaten bronze, sheet metal, wrought iron, gold, tree branches. Their inspired works mixed signs and allusions which called as much upon Pompeii as upon Egypt, Africa, and Oceania: "People invent extraordinary things for theater sets, why not show the same freedom with real furniture?"

A fairytale atmosphere of legends, which mixes the bizarre, the off-beat, and the asymmetrical on the one hand, and classical scrolls, Rococo mannerism, the curved lines of arabesques, and Napoleon III's stuffed boudoirs on the other. These are the imaginative trends of the designers who devoted themselves entirely to such decorative folly.

In London in 1986, the exhibition *Designers for Interiors* at the Victoria and Albert Museum established the neo-baroque around three great names: Mark Brazier-Jones, Tom Dixon, and André Dubreuil. Phantasmagoria, eclecticism, and post-nuclear catastrophe aesthetics play with revisited grand chic, ironic humor, and mad punk. Baroque 'n roll, as it is called by some, gained ground and became very hip. All the more so because mass-produced objects no longer come off the production line exactly identical: faultless products were not in fashion. Programming imperfection became terribly fashionable. And designers threw themselves into it wholeheartedly: Pucci di Rossi designed a chair wrapped in wire, François Azambourg a series of chairs in crumpled sheet metal that he called Mr. Bugatti in deference to the automobile manufacturer, Patricia Uriquiola a very haute-couture chaise longue, and Ron Arad the overly kitsch Ripple Chair dressed like a Moulin Rouge can-can dancer. The baroque style is theatrical and needs dramatic staging. Its entrance attracts attention.

The Sun King's pomp reinterpreted through this Zodiac mirror by Mark Brazier-Jones, 1986.

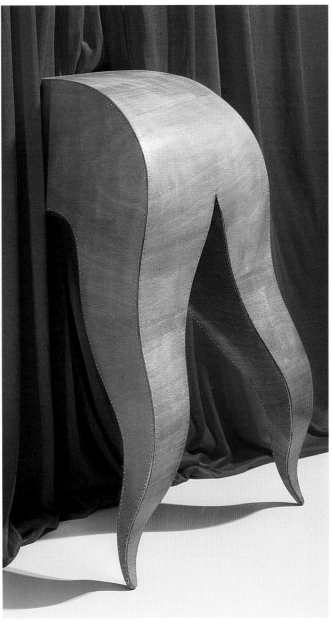

A cheeky pedestal table in precious wood, entitled Legs, designed and presented by Pucci di Rossi, 1990.

FOLLOWING PAGES Created by the designer Tord Boontje, these chair covers resemble ball gowns (Doll Chairs, Moroso, 2004).

are you MINIMALIST?

Contemporary art feeds your mind. You like the minimalist movement , vvhich appeared on the American scene around 1965, above all. Carl André's metal slabs on the ground, Donald Judd's rectangular structures, Dan Flavin's colored neon lights, or Sol LeVVitt's geometrical vvall paintings represent your idea of perfection. Looking backvvard, you applaud the vvords of the Austrian architect Adolf Loos in 1908: "Ornament is a crime." You subscribe unreservedly to the famous phrase "Less is more" launched by the German architect Mies van der Rohe in the 1930s. For you, the circle, the square, and the straight line contain all the beauty in the vvorld. You breathe better in the vvhite, open spaces of art galleries, perfect places for contemplation. Order and calm also reign at your home. You need solitude for meditation. It must be said, you are vvell read in Eastern and Zen philosophy, and you practice yoga vvhen you can. As for design, there is no doubt that the minimalist style corresponds to your idea of beauty.

*Could anything
be simpler or more
functional? Boxes
by Jasper Morrison
for Alessi (Tin Family,
1998).*

It is a tenacious movement that was not born yesterday: the experimental house presented at the Bauhaus exhibition in Weimar in 1923, and entirely furnished by the students, already attempted to make everyday life ultra-rational. With a special concern for facilitating the housewife's placing and disposition of objects. A new version of the minimalist style, closer to us, reappeared in its purest form in the early 1980s, when Japanese architects designed almost empty fashion stores.

This was the case of the Issey Miyake, Comme des Garçons, and Jil Sander brands, offering spaces in which the clothes were presented like artworks. In fact, this pared-down tendency referred to the radical ideals of the modern movement which, in the 1920s, wanted to make a clean sweep of the past and finish off once and for all the never-ending stylistic repetitions and other pastiches. It also appropriated the cult of the straight line and that of repetitive modules defended by the Italian group Superstudio which, in 1971, envisaged the creation of a group of neutral furniture pieces from a table and bench module called Quaderna in a parallelepiped form marked out in white squares. Out goes the decorative symbol. In come sliding walls, invisible cupboards, and elements to be built or fitted.

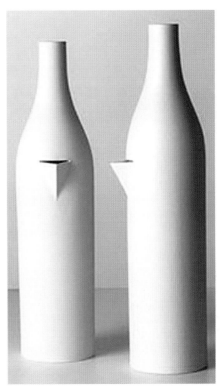

As elegant as they are practical, these immaculate white carafes rival minimalist sculpture. Carafe by the Bouroullec brothers, Torique collection, 1999.

The minimalist house was smooth, almost naked, freed of all that is unnecessary. Its almost Jansenist sobriety and rigorous rectilinear composition gave priority to the intellectual rather than the emotional. Rigor and measure reigned in this immaculate environment, bathed in an even light. The furniture was made from only the strict minimum.

Tables and chairs designed in straight lines combined formal simplicity and smooth surfaces. There was almost no color. Instead there were grays, off-whites, and creams. A few references are needed in order to obtain such sophisticated simplicity. It is important to know, for example, that modest intervention characterizes furniture pieces by artists or architects. It is indeed difficult to capture the elementary beauty of the famous Zig-Zag chair in a perfect Z shape, designed by Gerrit Thomas Rietveld in 1934, or the geometrical logic of the artist Donald Judd's furniture. The sculptor Bernar Venet also succeeded in creating beds, tables, and chairs whose sober elegance means we forget they are made of heavy metal. Among designers, the practice of "he who does less does more" style interests more than one. The simple contour of the Magis couch by Jean-Marie Massaud proves that there is no need to add more to make the essential a success. Others dream of transparency and immateriality, as Miss Blanche by Shiro Kuramata illustrates. Christian Biecher and his Slot couch, Arik Levy and his Slim double-shell chair, and Jasper Morrison with his Ply Chair Open chairs, one of which has an open back, the other a closed one—all practice a certain asceticism. Is the minimalist style discreet?

*The American
minimalist artist
Donald Judd
designed
elementary
and rectilinear
unit furniture
in the 1960s
(Desk Set, 1992,
JMG Gallery).*

FACING PAGE

*Designed in 1934 by
Gerrit Thomas
Rietveld, the Zig-Zag
cantilevered chair's
extreme simplicity
of line is impressive
(Cassina).*

are you NATURAL?

You are nostalgic for walks in the forest and strolls along the beach. You live surrounded by vegetation and objects evoking the marvels of the natural world. You favor authentic products, forms shaped by wind and water, and materials that demand to be touched. Your home exudes well being, calm, and serenity—but without selfishness. You contrast feelings of stress and urgency with daydreaming and contemplation. Saving the planet is one of your main concerns. As a child you spent a long time in trees, watching leaves rustle and listening to birds sing. These are vivid memories engraved in your mind, which you would like your everyday environment to reflect. The contemporary natural style suits you best.

What is particular about it? A specific way of combining raw materials with contemporary creation. The so-called natural aesthetic is characterized by a skillful dosing of bohemian simplicity, singular forms, and smooth textures. **It plays with daylight, tonal harmony, and surprising objects. Far from a strict extremist belief in craftwork, it combines the handmade with industrial fabrication. The need for nature at home is not related to any kind of militancy; there is no whiff of the hippie. Just a preference for certain materials. Especially wood.**

Roughly cut, it becomes a table, chair, or totem, while stone creates subtle arrangements. With the contemporary natural style, the artist becomes designer. Pierre Fuger, for example, makes original chairs from driftwood gathered along the Mediterranean coast and cleverly assembled in unstable positions. Christian Astuguevieille covers furniture with rope or hemp giving them a new appearance and a real sensuality. Blot Ken Wilson's shell mosaics make original décors that glisten in the light. With the contemporary natural style, cane, wicker, linen, and plant fiber, commonly used in the 1950s, come back in force. They rub shoulders with knotted vine, buffed pebbles, and grainy bark. Star designers take advantage of this to mix ancestral gestures such as weaving and stringing with the most specialized of techniques in a temporary association of the primitive with the rational. Marcel Wanders achieved this with his Knotted Chair, a chair made of a knotted sailor sweater in carbon fiber, as did Andrea Branzi when he designed his Animali Domestici chair which contrasts a backrest of rough branches fixed to a plastic structure. Or Hella Jongerius

Natural, light-colored wood for this airy ceiling lamp created by David Trubridge (Coral Light, 2004).

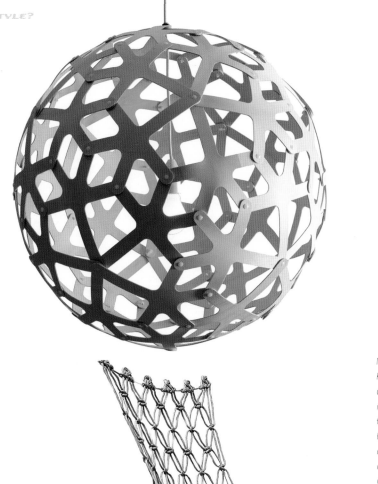

Marcel Wanders's Knotted Chair combines innovative materials and traditional weaving, high technology and wild-life dreams (1996, Cappellini).

when she makes industrial pottery such as Big White Pots that seem to come from an archaeological excavation site. However as far as a "shock" association is concerned, Jurgen Bey comes first with his Tree Trunk Bench, a tree trunk with old chair backs sticking out of it. David Trubridge goes back to the tradition of laminated wood in his group of light, perforated furniture pieces. **Certain designers make passing references to the art deco style by incorporating elements from the animal or plant world. The almost transparent sliding walls made of seaweed forms by the Bouroullec brothers or Fernando and Humberto Campana's Kaiman Jacaré divan bed with overlapping reptile cushions, shows the still-vibrant attraction of the wild. Domesticated or not, the contemporary natural style poeticizes our intimate spaces.**

copyright © ronan Bouroullec

This screen devised
by the Bouroullec
brothers brings
an air of poetry
to the everyday
world (Algues,
Vitra, 2004).

Many levels of
fabrication exist
between the tree
trunk and the chair.
Jurgen Bey's
surprising bench
offers a captivating
shortcut (Studio
Jurgen Bey,
Droog Design
for Oranienbaum,
1999).

IT'S ALL THE RAGE

IT'S ALL THE RAGE

TRANSFORMING FUNCTION

In art, reuse is a means of reinventing form, stripping down structure, or pushing back the limits of creation. It is also a means of introducing an often humorous distance from reality. Picasso excelled in this exercise by transforming a miniature 2CV automobile into a monkey head, and a bicycle seat and handlebars into a bull mask. Marcel Duchamp took the idea to its extreme by exhibiting objects bought in retail stores as artworks.

Contemporary design has taken possession of this practice which consists in reconsidering the object's function, form, and symbolism. Today, appearances can be deceptive: the Sassi rocks by Piero Gilardi resemble big, lifelike stones except that they are made of foam. Rob Brandt's crumpled white cups seem to be just like the office coffee machine's plastic ones, except that they are in fact made of china, and Marc Albert's lace lampshades are made of Sèvres biscuit porcelain.

Reuse also allows the designer to demolish notions of good or bad taste. Just by changing the object's usual field of operation he can create another understanding of the environment.

Mezzadro, a stool in the form of a tractor seat, is probably a way for Achille Castiglioni to bring the farm labourer's sweat into the wealthy suburbs, just as Stiletto makes a dig at bourgeois comfort with his Consumer Rest, a supermarket trolley transformed into a living room armchair. As for Matali Crasset's poufs and couch cushions covered in the same material as the Tati discount store shopping bags, they conjure up, wherever they are, the immigrant workers who, each summer, go back to their country using these bags as their suitcases. A change of scale is also a way of reconsidering the everyday environment.

As is the transferral from the useful to the agreeable, from the practical to the futile, from the function to the symbol. Gionatan De Pas succeeded in doing this when he devised the Joe armchair using a baseball glove as a starting point in homage to the American champion Joe DiMaggio, or when Gruppo Strum's work Patrone invites the consumer to throw himself down on a carpet of high grass.

Visual effects, a change of place, and swapping functions or fields all provoke a different understanding of everyday objects. And a smile!

FACING PAGE
Don't be fooled
by appearances:
these stones are
really light, soft
poufs (Sassi, Piero
Gilardi, Gufram,
1967).

Armchair shopping?
It's possible thanks
to Frank Schreiner's
association of
a shopping cart and
a chair for Stiletto
(Consumer's Rest,
prototype, 1983).

The urbanization of society: a tractor seat becomes a bar stool for fashionable interiors (Mezzadro, Achille and Pier Giacomo Castiglioni, Zanotta, 1957).

An original way
of converting
old coffee cups!
Or, how to reinvent
the banality of
everyday life
(5 Cups of Flower,
by Ron Gilad for
Designfenzider,
2000).

SALVAGING JUNK

A handful of people today live in opulence and overconsumption while the large majority is poverty stricken. Not everyone has the same access to everyday objects. One side throws away, while the other salvages. In the North as in the South precarity is spreading, and the poor have to be resourceful if they want to equip themselves: a drink can is easily transformed into a bedside light, an old door into a table, or a truck tire into an armchair. Charity organizations patch up, recycle, and rehabilitate abandoned furniture in order to give the poor decent living conditions.

Certain designers are aware of this reality and review their position accordingly. They turn salvaging into a creative process. This allows them to renew their style and assume a social role, relegating commercial objectives to the background. They can then create interesting and unique objects with a bizarre aesthetic, occasionally resembling sculptures or artistic installations, but which still remain functional.

The group 5.5 designers pushed this altruistic exploration the furthest and with the most fantasy. Its followers, all designers, created an original practice baptised "Réanim." Dressed in white coats, these men and women acted from time to time as doctors of objects. For a specific project, they followed a truck of a well-known charity association and collected the castoffs left by their owners. Then, thanks to a series of targeted "surgical" operations, with the help in particular of fluorescent green adhesive, ropes, plastics, and Plexiglas sheets, they managed to brighten up tables, chairs, and old armchairs. Humberto and Fernando Campana, who grew up in Sao Paulo, Brazil, are inspired by the craft techniques and objects made in shantytowns with whatever can be found. They make the most of each opportunity and create expressive, eye-catching furniture. Furniture like the Favela chair, an armchair made from pieces of wood, or the Banquete chair made from an astonishing assemblage of cuddly toys.

The creations of the Dutch group Droog Design are less exuberant but just as astonishing. Displaying an apparent distance from technology and industrial objectives, these designers breathe Dadaist poetry into everyday furniture and objects. They made, for example, a chest of drawers with old drawers found in flea markets, arranged in staggered rows and held together by a belt. In the same spirit, the famous vase by the Tsé 8 Tsé group juxtaposes several laboratory test tubes. As for Stuart Haygarth's spectacular light made of colored plastic scraps collected from a Kent beach, it is emblematic of the humor and generosity that characterizes the movement. **Recuperation? Rich with ideas.**

Making new out of old is the mission of the 5.5 designers group: Réanim project, from top to bottom: Joined for Life, table and chairs grafted together, 2003; Thonet Chair with Prosthetic Seat, 2004; Bistrot Chair with Crutches, 2004.

The Campana
brothers created
the Favela chair
from pieces of
salvaged wood
(Edra, 1991).

FACING PAGE
This spectacular
light is the result of
a lengthy collecting
of salvaged objects
(Tide chandelier,
designed and
produced in
10 copies by Stuart
Haygarth, 2005).

COMBINING MATERIALS

The history of design can be written through a history of materials. The designer today has both great freedom and a vast palette. He can use the oldest of natural materials, wood, and the most recent artificial fibers, as well as an infinite range of plastics and ultra-sophisticated metallic alloys. A catalog of varied procedures which includes craft techniques as well as specialized technologies is also at his disposal. Should he so wish, he can revisit his inventions with the benefit of hindsight. Occasionally transforming himself into a knowledgeable Professor Cosine. Several designers thereby experiment by willfully introducing breaks in the overly smooth logic of progress or by disturbing the rational chain of development for materials and techniques. How? Either by using hybridizations or material transplants, which were inconceivable up until now, technically or aesthetically.

Hella Jongerius is particularly fond of this process of making "entropic" productions which combine ancestral forms and new methods, or rather invulnerable materials and fragile textures. In this way, she manages to embroider china plates or, in a contemporary version of the flashlight, to juxtapose a camping stove bottle and a glass bowl. She also managed the impossible union of china and glass by creating two layered vases that she joins ironically with sticky paper inscribed with the word "fragile." Ron Gilad from the Designfenzider studio had the idea of designing a small, apparently precious, enameled china base. Placed on a common plastic water bottle, it plays with the contrast in materials in order to form an atypical one-flower vase. In a more masculine style, the American designer Ali Tayar experiments with the combination of glass, aluminum, and, for the first time, fiberboard, for his Nea table of rustic appearance paired with hi-tech methodology.

Only one step separates hybridization from the genetic mutation of forms, which the 5.5 designers happily take by creating garden benches and chairs whose bushy or shrubbery backrests demand to be pruned regularly. Enough to make you replace the cleaning lady with a gardener! Surprising isn't it, contemporary design?

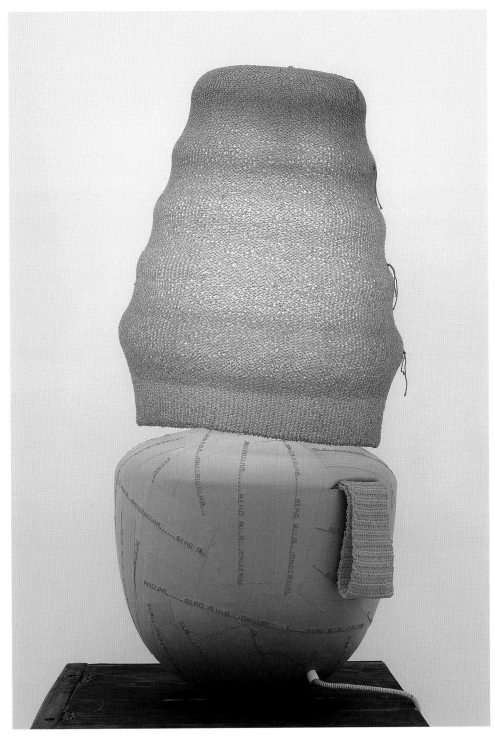

Two unexpected materials are juxtaposed here: knitted jersey and bandages for this bedside lamp put together by Hella Jongerius (Bead Bulbs, limited edition, Galerie Kreo, 2005).

This baroque
composition associates
volcanic stone and
laminated tabletop.
Élisabeth Garouste and
Mattia Bonetti graft
together, in this way,
the prehistoric and the
contemporary (Rocher
table, Néotu, 1983).

FACING PAGE
This independent
white unit can be
adapted to all
structures and
allows the user
to make his own
hybrid vase (Vase
Maker, Ron Gilad,
Designfenzider
Studio, 2000).

REVISITING THE PAST

A décor reveals the intimate tastes and personality of its designer. Not so long ago, the choice of antique furniture was the norm while opting for design was an exception. No dialogue was possible between the defenders of tradition and contemporary militants. Each stood their ground. Some praised the beauty of patina, the quality of finishing touches, and the permanence of forms. Some defended technical innovation, the renewal of materials, and a different conception of comfort. An invisible wall separated the traditional from the modern, thereby reactivating the break orchestrated in the 1920s by architects and innovators anxious to turn the page and to finish with hackneyed rehashings of the past. Postmodernism, marked by the arrival on the Italian scene in 1979 of Studio Alchimia and the great return of baroque forms, buried the hatchet. The young go-getters of design now look at furniture from the past in a new way. Without feeling as if they were letting the side down.

They take unreservedly from the great antique shop represented by the history of styles. They update cult forms, with proportion or disproportion.

Gently, as Daniel Rozensztroch does, one of the first decorator–designers to incite readers of interior design magazines to take liberties with grandma's Regency chest of drawers or Uncle Paul's copy of a Biedermeier chair. How? By painting the chest of drawers with—why not?—a layer of red lacquer and by covering the latter with an African fabric. Fashion designers such as Jean-Charles de Castelbajac and Maurice Renoma also redesigned the famous medallion chair, one with flashy colors, the other with photographic prints. Emma Roux replaces the backrest or seat of a Louis XV cabriole chair with translucent, colored Plexiglas. Building bridges between yesterday and today can create even more surprising results. Philippe Starck makes a successful takeover bid with his "grand style" chair in translucent plastic. Alessandro Mendini from Studio Alchimia, on the contrary, is more theatrical and baroque and lays superfluous decorum on more thickly with his arm-chair–throne Poltrona Di Proust. He confronts a symbol of bourgeois power with the free and sensitive brushstrokes of the painter Seurat. Much less sensual but just as spectacular is Stefan Zwicky's homage to Le Corbusier, his concrete version of an emblematic chair from the early twentieth century.

Two surprising interventions pay a tribute to history.

A rickety armchair by the designer Jurgen Bey, (Healing chair, collection Droog Design, 2000).

Burnt chair by Maarten Baas (Smoke Collection, Hill House, Moss, 2004).

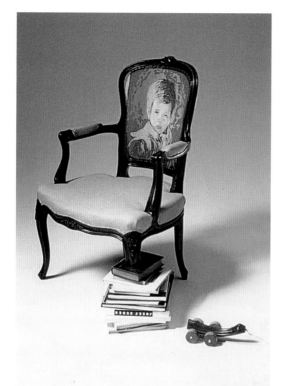

As for Pablo Reinoso, he subjects the famous Thonet chair to all possible contortions. Other designers create pieces hovering on the edges of art. Jurgen Bey from the Droog Design group wraps a series of antique chairs in thin plastic film. His Kokon functions like the ghost of our memory. He also plays with the idea of past time and culturally loaded antique forms by amputating the leg of a Regency armchair and replacing it with a pile of books. Even more of a transgression is Maarten Baas's burnt furniture, which is shown half carbonized. The effect is striking.

There are no longer any barriers between yesterday and today, but rather a game of distorting mirrors.

The artist Pablo
Reinoso makes
sculptures which
subject the Thonet
brothers' iconic
chair to all sorts of
transformations.
(2005).

FACING PAGE
Stefan Zwicky creates
a concrete version
of Le Corbusier's
famous armchair
(Dommage à
Corbu, 1980).

BLENDING STYLES

Fashions pass, movements follow on from each other, trends intersect. It was ever thus. Today, however, the phenomenon has gathered pace. Global zapping makes our head spin with its kaleidoscopic image of a jigsaw-puzzle world. The furthest lands come together while identity-forging customs find a new lease of life. Internet and the mobile telephone dissociate space and time. We can be both everywhere and nowhere at the same time: we write blogs, chat with strangers, exchange different viewpoints, and discover other ways of living.

How can we resist this amazing effervescence and keep to our usual habits? How can we not feel concerned by this incredible telescoping of information? How can we ignore this wealth of new exchanges?

At the beginning of the third millennium, many contemporary furniture or interior designers work on translating this extraordinary flux by practicing a form of stylistic sampling. Like DJs, they force a cohabitation between cultures and periods, mixing the old with the new, high design with tribal art, and the east with the west. They gladly flirt with contradictions and bad taste and succeed in creating daring combinations which horrify the traditionalists.

In this way, they dare to juxtapose a constructivist armchair by Gerrit Thomas Rietveld with a Murano chandelier, or make unnatural associations by exhibiting a Fang mask in front of a Rococo tapestry, or bringing together the curves of a Charles and Ray Eames rocking chair with a rigid Mackintosh armchair. They marry inlaid marquetry with linoleum, gold or silver plate with plastic, and shantung with fake fur, to the great displeasure of the guardians of good harmonious taste. Achieving successful alliances has become their aim, the reconciliation of antagonisms their credo. Shuffling the icons of art deco, Bauhaus, Scandinavian design such as Ikea, and Pop, the most talented invent an eclectic style that questions the usual rules of harmony and elegance. By joyfully scrambling rituals and influences, they develop an exoticism which defies notions of time and latitude. As far as objects are concerned, a few designers attempt to build bridges between the past and the present. For example the Capitello stool by Studio 65, which represents the broken top of a column from the ruins of a Greek temple, or Massimiliano Adami and his prototype piece of furniture Fossile which makes shelves out of prints of common, universally familiar objects such as the television, the pack of mineral water bottles, and the transistor radio.

Yesterday, today, and tomorrow are finally reunited.

This famous living room, with both contemporary and eighteenth-century furniture, is in the Haussmanian apartment of a Parisian decorator (German armchair, luminous coffee table, and Italian chandelier).

The designer
Jurgen Bey poses
in the middle of
his creative jumble
(The Dust Cabin,
Jurgen Bey studio,
assisted by Henriette
Waal, 2005).

GREEN HORIZON

VVe may as vvell prepare ourselves for the idea that frenzied consumption and vvaste vvill soon belong to a past era. The statistics on sustained development announced by the vvorld summit held in Johannesburg in 2002 speak for themselves: if current trends continue, in 2050 the vvorld's population, vvhich today nears 6 billion, vvill be 3 billion more. And if vve do not change our consumer habits, vve vvill need three more planets to continue living as vve do. All the more so because the recent economic grovvth of developing countries only increases the earth's ecological problems: the greenhouse effect, ozone holes, air pollution, vvater loss, the disappearance of biodiversity, and so on. Added to all this is the vvorrying problem of the recycling of vvaste, due to the over-production of objects.

Having long been keen follovvers of progress, designers today are vvorried and rally in support of ecology. Situated at the crossroads betvveen business and the consumer, they have started to consider their role differently. They vvant to participate actively in a change of outlook by trying to create other habits, other attitudes, and other expectations tovvard the objects that surround us. **Their aim? To change profoundly the way in which wealthy societies produce and consume. But how to go about doing this? At first with the help of simple means: by reducing the weight and the volume of objects, diminishing their energy consumption, using recycled materials.** The following fevv examples exist in this virtuous category: a brand of athletic vvear made out of recycled plastic bottles (Patagonia), a backpack vvith integrated solar panels for using an iPod, a mobile

An ecological radio
(EyeMax, Freeplay,
2005).

140

telephone, or even a camera (O'Neill), and a solar radio which can be recharged with a crank (Freeplay). From now on, products benefit from a good reputation if they resemble the simple, essential products that can be found in the mail order catalog created by Philippe Starck for La Redoute in 1999. Or like the uniformly monochrome, sober, and discreet supplies sold by the Japanese company Muji, which testify to both a humble and economical spirit.

Designers need to be even more reactive however in order to satisfy current demands without compromising the needs of future generations. They have to take into account environmental objectives without limiting themselves to just the customer, but extending their understanding to all those affected directly or indirectly by objects and their uses. Why not rethink then the very notion of product?

Small tables in recycled cork (Jasper Morrison, Cork Family, Vitra, 2004).

The experiment conducted in France in 2002 by the design agency O2, entirely devoted to ecological design, undertaken in collaboration with WWF France and Victoires Éditions, shows the way. It reflects upon the relationship we have with the products we use to satisfy our essential needs. The idea being to push consumption habits toward services. This experiment resulted in the proposal of alternative scenarios which invite us to adopt a more responsible, more generous, and more collective attitude. Maybe one day we will see a fruit squeezer and a shaker at our disposal at the supermarket, a foldable bike and child seat rental service at the parking lot, and a furniture repurchase, repair, and reupholster service in DIY shops. Each room in the house will probably have an energy meter, allowing us to see the electrical consumption of each appliance, while in the kitchen, the sink will be equipped with a special tap giving tap water the properties of mineral water. As for the slippers we slip into when we come home in the evening, they will have piezo-electrical soles capable of providing energy for an MP3 player or a portable computer. There is no time to lose.

This jacket is made out of recycled material (Recycled Jacket, M's Synchilla Snap T, 1993).

THE NOMADIC SPIRIT

The global economy stimulates more movement. Revolutions in telecommunications transform us into nomadic travelers: equipped with a portable telephone and computer, we enjoy an unprecedented freedom of movement. In return, we have to be both adaptable and reactive. Let's prepare ourselves for moving home and changing jobs in the future. What relationship will we have with the home in the future? Anxious to reflect the contemporary nomadic spirit, designers are interested in the fact that today the barriers between domestic and workspace are blurred.

Observation, estimation, and forecasting have gathered pace. Contradictory myths of the mobile home overlap. It is both a symbol of adventure and a burrow for protection, withdrawal, and a place to be alone. Trapped between archaism and new technologies, the postmodern designer swings from one domain to another, offering radically different solutions according to the day and the hour. Monica Föster's mobile space Cloud, which can be installed inside the home, reflects the need to get back in touch with the inner self, whereas the backpack armchair by Olivier Peyricot prompts an urgent need to travel and get out of one's shell. **Each object must also be practical and lightweight. However a flexible object is also unstable. Increasingly aware of the disparities and living conditions of those left behind, designers think in terms of solidarity and urgency.**

A few designers work on the problem of building shelters or emergency refuges in urban areas. Inflatable structures are often the best solution for this. Michael Rakowitz's ParaSITE is a clever survival tent for the homeless which is inflated and warmed by a pipe which recycles air conditioning from the surrounding buildings. Similarly, Martin Ruiz de Azúa created a sensation in 1999 with his big cube La Casa Basica, the prototype of an inflatable home made out of the same material as survival blankets. It is pocket size but when blown up, it becomes a tent. It can also be used as a protective cover thanks to its insulating properties.

Inflatable objects and furniture are also popular for more joyful occasions. Immediately ready for use, inflatable products can adapt to impromptu events and circumstances: Fred Greneron's inflatable Patapoof bar, and Nick Crosbie's inflatable fruit bowls, eggcups, and vases bring poetry into an overly rational world. Movement's fine, but gently does it.

This inflatable vase was designed by Nick Crosbie (Starvase, Inflate, United Kingdom, 1996).

The inflatable Patapoof
bar by Fred Greneron
allows the organization
of spontaneous parties
(Du Bonheur!, March 2003).

La Casa Basica by
Martin Ruiz de Azúa
is pocket-sized.
It can adapt to
all emergency
situations (made
by Pai Thio, 1999).

FUTURISTIC OPTIONS

Are we still capable of seeing into the future at a time when Internet and the computer have upset our relationship with time and space once and for all? "Feet on the ground can be found here," states provocatively Enzo Mari, the high priest of Italian design. The fact is that now even the craziest dreams can take shape almost immediately. And virtual reality sometimes goes beyond fiction. It remains for designers to invent tomorrow's and the day after tomorrow's objects. Realistic or not, there will be surprises: everything seems possible now.

TOMORROW IS NOW)

The Ubiquitous Media Chip Gumi (*sic*) perfected by the research center of the Japanese manufacturer NEC flirt, in this way, with science fiction. Sold as candies, these transparent capsules containing micro-chips can be swallowed once the images and the music they contain have been used. As for the little Qrio dog created by Sony Corporation, will it replace Rover or Spot one day? Measuring 22in. (58cm), it is sweet and cuddly. It can jump, avoid obstacles, recognize faces, and understand two thousand words. Will we still be wearing clothes in ten or twenty years? Will intelligent fabrics function like a second skin? X-Technology has come up with X-bionic underwear which resembles a diving suit but affords better circulation of the blood during sporting activities. There are also anti-perspiration, anti-bacterial, and healing fabrics. Others adapt to temperature range, or resist stains and discoloration. The home, as an ideal place for experimentation, will undergo great upheavals. In interior decoration, the patterns and colors of fabrics will be variable thanks to preset programs. Electronic networks integrated into the heating, lighting,

and alarm systems will make it possible to control everything from the computer.

The e-home means the consumer will be able to try out the new ways of defining "comfort" and "interactivity." An intelligent home will lower the heating of an empty room, detect the end of the mineral water reserve in the refrigerator and order more, or reduce human and food waste to dust. Cleaning equipment will respond to vocal commands, communicate between separate units, and have several functions. Furniture will be almost non-existent, replaced by evolving structures which adapt to different domestic rituals, similar to those of Karim Rashid or Frédéric Ruyant's radical proposals.

Likewise, Igor Novitzki's biomorphic lights seem to float in an undetermined space. Finally and even more impressively, bodily implants encoded with personalized information could replace keys and other access codes to entrance doors. Home automation excites designers who consider it the new El Dorado of utopian design.

FACING PAGE

This otherwordly object is in fact a light (Drosephylia, Igor Novitzki and Yann Guidon, Natural/Digital exhibition, galerie numeriscausa, 2005).

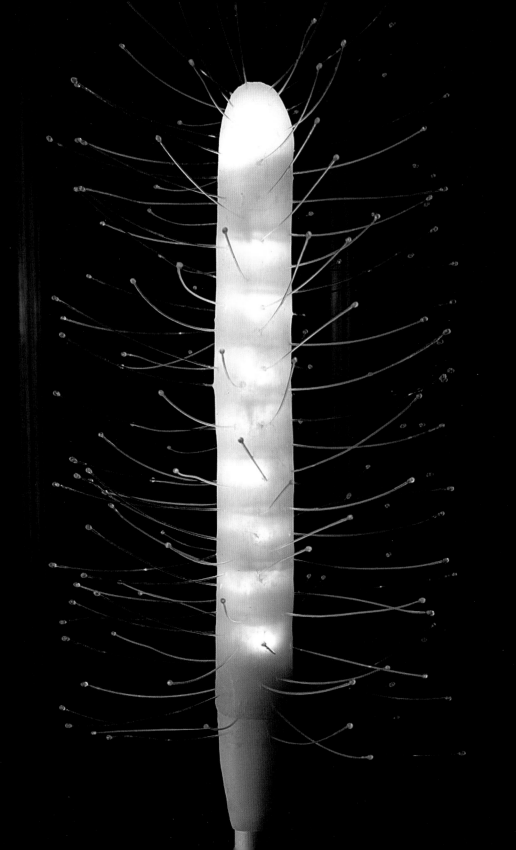

Tomorrow is already here! The capsules in these tubes are sold like candies and contain digital information (Ubiquitous Media Chip Gumi, NEC Corporation 8 NEC Design, 2003–04).

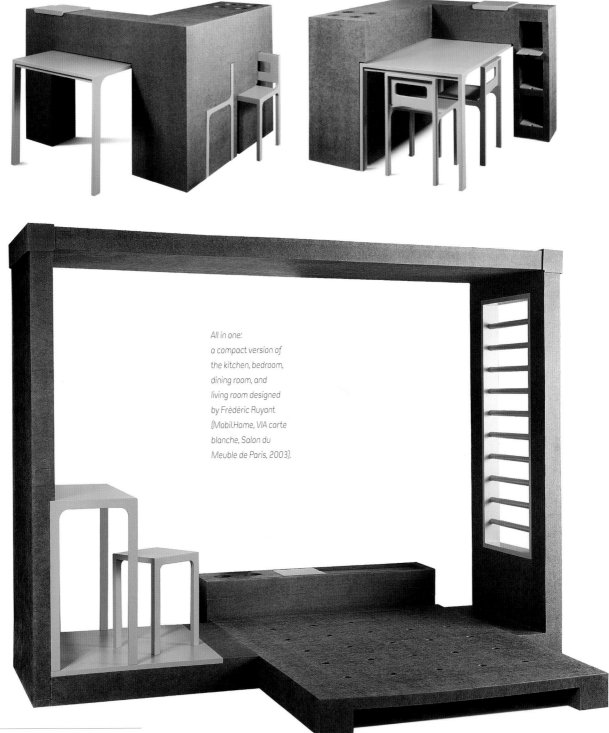

All in one:
a compact version of
the kitchen, bedroom,
dining room, and
living room designed
by Frédéric Ruyant
(Mobil.Home, VIA carte
blanche, Salon du
Meuble de Paris, 2003).

"MULTIFUNCTIONAL"
ORIENTATION

Not so long ago, when we got married, it was for life. The bedroom, with its double bed, two bedside tables, a beautiful wardrobe, and a comfortable walnut or chestnut chest of drawers received as wedding present, represented a lasting relationship. In a few decades, the sexual revolution and the valorization of the individual have overturned this pattern. Blended families or single parents are obliged to adapt to home models conceived according to lapsed norms. Spaces and furniture designed according to a traditional schema hardly ever take into account different ways of living. In order to accompany these new practices, designers today therefore imagine other kinds of relationships with the objects that surround us.

They are already projecting themselves into a future where furniture will have several functions and where spaces are polyvalent. There is a strong, very realistic chance that the famous trio of a couch, coffee table, and two armchairs will undergo various metamorphoses. Prospective projects tend toward multiple shapes.
A piece of furniture can now hide one, two, or three others, and more.
    Pierre Charpin's elegant couch-armchair-chaise longue, composed of stacked modules, the number of which determines the length, is a successful example: it responds to all needs. Another successful module is Pierre Bauchet's corner couch which, like a puzzle, stacks one armchair upon another, although they can also be used separately.

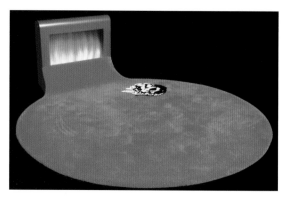

A relaxation space presented as a ready-to-use décor (Sleeping Cat, by the Radi Designers, Kreo/Radi Designers, limited edition of eight, 1999).

Christophe Marchand's Matrable is even more surprising. It is a kind of compact block which includes a table, a chair, and a resting space in its stiff foam mass. Foldable and unfoldable furniture also plays with new forms, such as Martin Szekely's metal wardrobe which at first resembles a flat metal sheet. It is then set up by closing the spaces indicated by the dots. Azumi's S8T table is even more interchangeable as it can be articulated according to various uses: high table, coffee table, cupboard, shelves, etc. Improbable forms for indefinite positions. This is what comes to mind when faced with the Sleeping Cat rug by the Radi Designers which invites us to sit or lie down next to a false fireplace drawn as an extension of the rug itself, or the Petal armchair by Ora-ïto which resembles neither an armchair nor a chaise longue, nor a couch. And what can be said about Konstantin Grcic's hybrid three-legged seat which serves as chair, stool, and writing table according to whether we sit sidesaddle or astride it? It adapts to different uses for different situations. Furniture in tune with its age!

Chair, table, and stool: this seat combines different functions (Allievo, Konstantin Grcic, Montina, 2000).

THEY DID IT FIRST

THEY DID IT FIRST

A FRONT ROW SEAT
TO INNOVATION

There is nothing more ordinary than a chair. Hundreds and hundreds of different models exist, however. This basic component has given rise to numerous interpretations. In the same way as a pianist practices his scales, the designer designs a chair or armchair. Both are an excellent exercise in style. For certain designers, the chair is a manifesto of sorts. This is why, when it comes to furniture, all great designers dream of making their mark with an armchair or a chair that will symbolize a decade or a period. Yet, when we look at modern and contemporary history, the great stylistic turning points are closely linked to technical innovations. And the stories of these "first times" show how seldom a new form ever appears by chance.

Here is the story of this incredible chain of cause and effect represented by iconic chairs. We can see how ideas bounce off one designer to another. And how those who innovated and created original forms are the authors of models that today are considered design classics.

A humorous hommage to design! This chair inspired by Mallet-Stevens is made with inflated balloons (Pink Chair, Michel Mallard, limited edition of seven, 2006).

154

1859

The Thonet brothers
Chair model no. 14

1925

Marcel Breuer
Model B3, Wassily armchair

The modernists dreamed of it, the Thonet brothers had already done it!

Created in their factory at Koritschen in Moldavia from 1859 onward, Model no. 14 fulfilled the dream of the innovators of the late nineteenth century: to offer beautiful, cheap mass-produced furniture. As a symbol of the shift from the craft industry to industrial production, this mythical model owes its existence to the perfection of the technique of bending wood with steam. Constructed of pieces of wood shaped in this way and held together with screws, it would serve as a base for a range of varied products all made along the same lines, and delivered in a kit. It was met with popular success: by 1911 fifty million units had already been sold worldwide.

A German architect and designer of Hungarian origin, Marcel Breuer's Wassily armchair was the first chair with a chrome-steel tube structure. This creation first saw the light at the Bauhaus school of applied arts in Dessau, where Breuer directed the furniture workshop. He had the idea, revolutionary at the time, of bringing together the cabinetmaking workshop and the metal workshop to carry out research on new techniques. Until then, the metallic tube had only been used for bicycle frames and airplane structures, utilizing a process that avoided any joins. It had never been used for furniture. The Wassily, named by Breuer in homage to the Russian constructivist artist Wassily Kandinsky, was an important innovation. However, its design meant it could not be produced industrially. The model was perfected two years later by the architect Mies van der Rohe, who also taught at the Bauhaus, and it met with the same difficulties. But the fashion was launched: in the 1930s, a vast range of seats, tables, and desks in metallic tubes, produced and distributed on a large scale by the prosperous Thonet brothers' company, met with great success.

1926

Mart Stam
S33

1931

Alvar Aalto
The "Paimio armchair"

Before Mart Stam created this cantilevered chair, an armchair or a chair had legs. The Dutch architect, who belonged to the radical and committed avant-garde, taught at the Bauhaus with Marcel Breuer when the latter designed his famous Wassily armchair. Keen to experiment with the technical possibilities of metal, he succeeded in making a first chair. By testing steel's resistance properties, he managed to perfect a cantilevered chair with an innovative outline in 1926: it seems to be floating in the air. Other designers, with other materials, would later take this fascinating lesson in balance further.

This gently curved seat was designed by the Finnish architect Alvar Aalto at the time of his construction of the Paimio sanatorium. This Paimio model no. 41 postdates by a few months the Springleaf model no. 379 in which Aalto used bent plywood for the first time. However the Paimio, in curved wood lamellate, became a furniture icon as it is considered the first armchair of organic design. And if its form adopted the blueprint of international style, it nevertheless broke with modernism's obsession with geometrics. Here, everything is about harmony, comfort, and conviviality. The back and seat are part of the same piece giving it a certain homogeneity, their scrolled extremities ensuring more flexibility. In this way, the Paimio offers the same comfort as a club chair.

1948

Eero Saarinen
The Womb Chair

This enveloping chair-armchair, conceived by the Finnish sculptor and architect Eero Saarinen, responds to the challenge put forth by the avant-garde furniture designer Florence Knoll to create "a chair she could curl up in, or roll up in." Technically, it is a first as its seat is made of a fiberglass shell covered in fabric and placed on a polished chrome steel rod. This model is one of the great classics which gave the brand its fame.

1950

Charles and Ray Eames
DAR armchair

Made of molded fiberglass and reinforced polyester, the Dining Armchair Rod (DAR) armchair was a revolutionary concept in seating. A material previously only used in aeronautics, to reinforce the domes of airplane radars, was applied to furniture for the first time. Just as it is, bare of all ornament or covering. The machine finishing of polyester resin reinforced with fiberglass had now been totally perfected.

The fact that Charles Eames, who trained as an engineer, had worked for the army during the Second World War was significant. At the end of the war, Charles and his wife Ray conceived of this reinforced and molded polyester with the aim of creating a lightweight seat made of only one material without screws, rivets, or fitting together. They then produced other versions of it with different stands.

1954

Osvaldo Borsani
P40 chaise longue

The P40 chaise longue was the first to use another material taken from the automobile industry as stuffing—latex foam. It was designed by the Italian architect Osvaldo Borsani from the Milan Technology Institute, who, in 1953, founded the Tecno design company. This reference model can be adjusted to forty different positions. The head-rest, backrest, seat, and foot-rest are independent. Its pared-down style owes its particularity and its modernity to Borsani's long-term research on technique. He was, moreover, the founder of the review *Ottagono*.

1956

Eero Saarinen
Tulip chair

In 1940, Eero Saarinen and Charles Eames had each won, separately, the Organic Design Home Furnishing competition set up by Eliot Noyes, the director of the MoMA Design Department. Their proposals took the shape of plywood shell structures, covered in foam and fabric. They have harmonious but industrially inexploitable forms. In 1953, Charles and Ray Eames moved a step ahead with their production of the first mass-produced plastic shell seat. Aggrieved, Eero Saarinen countered with his concept of a seat molded in one single piece. Despite numerous technical attempts, he never succeeded. However, his Tulip chair, perfected in 1956, might lead us to believe otherwise: its metallic stand is the same color as the seat, so it seems to be homogenous. And it was the first successful attempt in creating a continuation between the legs and the shell of a chair.

1963

Pierre Paulin
Mushroom Chair F560

The first of its kind, this rounded, textured armchair was made only after numerous technical attempts. It is emblematic of the gentle, fluid lines which brought the famous French designer Pierre Paulin international acclaim. Made of a metallic structure covered by seamless elastic wool crepe which is pulled on like a sock, it is soft and comfortable.

1963

Peter Murdoch
Spotty child's chair

In the 1960s everything was possible. Unconventional and democratic, the period inspired designers to make common consumer products which broke with the idea of heritage and transmission. Ephemeral furniture appeared, successfully embodied by the first foldable and un-foldable cardboard chair: Spotty. Created by the British designer Peter Murdoch, this coated cardboard seat was sold flat at a modest sum. It was the first of its kind to be marketed, and had a lifespan of six months. The colored spotty pattern is reminiscent of pop art's carefree and joyful atmosphere.

1964

Marco Zanuso and Richard Sapper
Model K 4999

In the 1960s, the Italian firm Kartell specialized in the production of plastic furniture, commissioning the best designers. It invited them to explore the possibilities offered by this material. A graduate of the Milan Polytechnic, the architect and designer Marco Zanuso devised children's chairs that were the first to exploit the properties of polyethylene. This lightweight material has the advantage of being easy to handle. Nevertheless, the firm took five years to undertake the mass production of the K4999 model. These chairs, which could be stacked like Lego, were immediately successful in kindergartens: amusing, light, and brightly colored, they could be used by children to build cabins and play areas.

1964

Helmut Bätzner
BA 1171 chair

Launched at the furniture fair in Cologne, this chair was created by the designer Helmut Bätzner. Made of resin and reinforced fiberglass, it was the first synthetic chair to be suitable for mass production. Using the "Prepreg Process," it took five minutes to make, and the finishing touches not much longer. It was met with commercial success: its sober aestheticism and good quality–price ratio made it unbeatable.

1967

G. De Pas, P. Lomazzi, D. d'Urbino, and C. Scolari　**Blow armchair**

These four Italian designers devised the first marketable inflatable armchair. Others before them, notably Verner Panton, did not get beyond the prototype. Gionatan De Pas, Paolo Lomazzi, Donato d'Urbino, and Carlo Scolari were able to go further by taking advantage of advances in plastics technology, using a technique of radio-frequency soldering also known as PVC electronic soldering to seal the seams. The Blow armchair is sold flat like a deflated rubber ring, at a modest price. Its sober lines recall the modernist aesthetic, but its acidulous colors are in keeping with the spirit of pop art.

1968

Pierre Gatti, Cesare Palini, and Franco Teodoro　**Sacco**

Symbolic of the rebellious and non-conformist spirit of 1968, the Sacco reinvented the chair. With no legs, seat, or back, the leather or vinyl bag full of millions of expanded polystyrene beads weighs less than 14 lb (6 kg), and molds to the shape of the body as it sits. This first soft, completely ergonomic chair invited voluptuous laziness, responding to the desires of a generation who wanted to lie on the floor and listen to floating music for hours on end. Amazingly comfortable, it was perfectly suited to the period's moral revolution.

1968

Verner Panton
Panton chair

1969

Gaetano Pesce
Up seats series

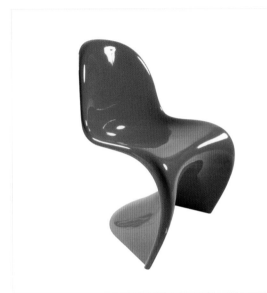

Designed in 1960, the famous Panton did not go into production until a few years later. Its perfection required long technical research. Verner Panton originally dreamed of realizing the aim of 1950s designers like Charles Eames and Eero Saarinen who wished to create a shell chair out of a single block. Inspired by this challenge and by the wooden zigzag chair devised before the war by Rietveld, Panton designed the S chair in 1960. But he wanted more: to make a stackable object of dyed plastic that was smooth, glossy, and capable of resisting extreme pressure. With the help of engineer Herman Miller he was able to take advantage of the development of sufficiently resistant thermoplastic materials to accomplish this feat.

Greeted with "ahs!" and "oohs!" when it was presented at the Milan furniture trade show in 1969, the Up seats series, signed by the Italian designer–architect Gaetano Pesce, was a big hit. Such innovation! Sold in a vacuum pack, these armchairs, poufs, and polyurethane cellular foam couches covered in extensible jersey fabric inflate to their original form as if by magic when unwrapped. Pesce's masterstroke was to take the air out of the foam cells once the seats had been formed, thereby reducing their volume to a flat pancake. Wrap it in a PVC envelope and the game was won!

1972

Frank O. Gehry
Wiggle Side chair

The Canadian–American architect Frank O. Gehry likes to test unusual materials. He was the first to come up with undulated cardboard seats, giving particular attention to their form. Both extremely robust and agreeable, the Wiggle Side chair resembled no other existing chair. Its undulating lines play on the contrast between flexibility and rigidity, giving the chair a sculptural appearance.

1987

Alberto Meda
Light Light chair

Only fifty copies of this carbon fiber chair exist. It was made as an experiment, and its anthropomorphic silhouette has a futuristic air. Previously carbon fiber had been used in sports material where its exceptional lightness and resistance were appreciated. Due to the high cost of manufacturing, this seat was never commercialized. A shame, as its elegance and lightness (it weighs less than 2 lb/1 kg) are considerable assets.

1998

Philippe Starck
La Marie chair

The La Marie chair is totally transparent and is made from one single molded polycarbonate piece. Unlike Plexiglas, it does not scratch. Considered the first invisible chair, it belongs to the list of "good" objects dreamed up by Philippe Starck. As he says himself, it is "the indispensable non-product." It figures as such on page 51 of the Good Goods mail order catalog created by Philippe Starck for the La Redoute company. La Marie is the anti-design object par excellence. It melts into the background.

2004

Patrick Jouin
Solid C1 seat

This seat would not exist without the use of a rapid prototyping process called stereolithography. The designer Patrick Jouin designed this seat precisely in reference to this process. Commonly used for the making of models and synthetic materials, it considerably reduces manufacturing time. This chair took shape through the superposition of thin layers of liquid resin which solidify as they cool. This procedure is repeated until the programmed form is obtained.

30 DESIGNERS and their signature works

30 DESIGNERS and their signature works

RON ARAD

who is he?

The man with the hat was born in Tel Aviv in 1951. He studied architecture at the Architectural Association in London where he was taught by Peter Cook and Bernard Tschumi. He founded the design studio One Off Ltd. with Caroline Thorman in London in 1981. Eight years later, he set up on his own as Ron Arad Associates. Designer, architect, stage designer, and film director, the prolific Arad constantly reinvents the grammar and syntax of forms, daring to outrage popular taste and to question function.

his style

Curves, spirals, and scrolls characterize the futuristic contours of Ron Arad's furniture. Their rounded forms are often countered by a choice of masculine materials such as steel or stainless steel. Shiny or opaque, rough or refined, light or dense, this researcher's productions are inspired by new materials and boast sumptuous sculptural forms. They move with apparent ease from the stage of unique prototype to mass production. However, their perfecting, which is often complex, reveals the designer's passion for innovative technical processes. His Ripple chair, which copies the intertwining pattern of the Möbius strip; the Un Cut Chair, which takes its shape from a print on a metal sheet; the famous spiral Bookworm shelves; and the voluminous Big Soft Easy chair have all found a place in the pantheon of contemporary design classics.

best of

1994 Empty chair, Driade ❖ **1995** Interior architecture of the Adidas Stadium in Paris ❖ **1997** Un Cut Chair ❖ **1998** Tom Vac seat ❖ **1998** Fantastic Plastic Elastic chair, Kartell ❖ **2004** The MT seat collection, Driade ❖ **2004** Lo-rez-dolores-tabula-rasa installation at the Milan furniture fair and the Venice Biennale ❖ **2005** Miss Hazé lamp for Swarovski

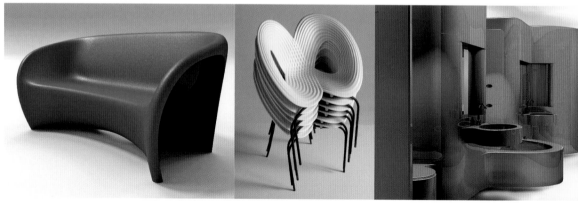

MT Sofa, Driade, 2005.

Ripple Chair,
Moroso, 2005.

Puerta América Hotel, 2005.

Soft Big Easy, steel structure, foam, and synthetic padding, covered in pure untreated red wool, Moroso, spring collection,1991.

He said:

"Design's fight against received ideas can be liberating."

RONAN AND ERWAN BOUROULLEC

who are they?

Elder brother Ronan graduated from the École Nationale des Arts Décoratifs in Paris. He began his career as a designer in 1997 by designing combinatorial vases for the Galerie Neotu in Paris, followed by a disintegrated kitchen for the design agency VIA. His brother Erwan, a graduate of the Cergy-Pontoise art school, joined him in 1998. Since then, the young Breton prodigies have designed production after production, non stop. Over the years they have gained the status of leaders of a new, inventive and poetic French design. They are in demand by major international museums and work for the biggest design producers, including Vitra, Cappellini, Magis, Ligne Roset, Habitat, and Kreo.

their style

What relation is there between a perforated partition woven out of fine plastic strands in the form of seaweed, an ultra-thin desk-table, and shelves placed on top of endlessly stackable white cells? The pleasant feeling of floating in the air, of stepping back from a sometimes burdensome reality, and of entering a protected zone. Adjustable and transformable, the flexible shapes devised by the famous duo connect and disconnect privileged relationships with their users. And follow the random curves of our mobile lifestyles.

best of

1998 Cuisine Désintégrée, Cappellini ✣ **2000** Lit Clos, Cappellini ✣ **2000** Design of Issey Miyake's new clothes space A-Poc, Paris ✣ **2001** Outdoor chair, Ligne Roset ✣ **2002** Exhibition at the Design Museum in London ✣ **2002** Joyn desk, Vitra ✣ **2005** Elected designers of the year by the Salon du Meuble, Paris

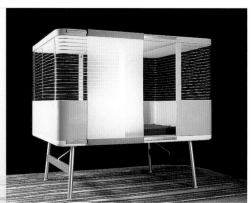

Facett chair and Bridge,
Ligne Roset, 2006.

Honda vase, Galerie Kreo, 2001.

Lit clos, Galerie Kreo, 2000.

Shelf Shelf, ABS polycarbonate, Vitra, Switzerland, 2004.

They said:
"Design is about knowing how to analyze both the situation and the partner with which we work."

HUMBERTO AND FERNANDO CAMPANA

who are they?

The "enfants terribles" of South American design were born in Sao Paulo, Brazil: Humberto in 1953, Fernando in 1961. The elder of the two studied law at university, the younger studied architecture at the school of fine arts. They became designers on the job: in 1983, Fernando helped his brother Humberto who had embarked upon sculpture, to make metal furniture pieces, and has stayed on ever since. Since 1991, when Favela, an armchair made out of salvaged wood was created, the Campana brothers' eccentric creations have never ceased to surprise us.

their style

The Campana brothers came up with an explosive version of the North–South encounter: modern lines but humble materials. Exuberance, vitality, and fantasy define their baroque, almost kitsch, style. Their abundant creativity combines shantytown resourcefulness with a spirit of reuse, a key concept of contemporary art. For them a broom becomes a chair back, fabric samples become a pouf cover, and old tangled ropes become the seat of an armchair. But there is only one step between amateur recycling and the market economy: their furniture pieces are produced by Edra and enchant the happy few of the planet. They are already in the collections of important museums.

best of

1991 Favela chair, Edra since 2003 ✲ **1999** Jenette chair, Edra ✲ **2000** Anemone chair ✲ **2002** Banquete chair ✲ Sushi chair, Edra ✲ **2004** Alligator chair ✲ **2006** Kaiman Jacaré, Edra (prototype)

Kaiman Jacaré, Historia Naturalis collection, Edra, 2006.
Launched in 2006 at the Milan furniture fair.

Banquete chair, Studio Campana, 2002.

Sushi IV, Studio Campana, 2003.

They said:
"The streets of São Paulo are a kind of laboratory for our ideas. Whenever we need inspiration, we rely on the chaos of the city in which we live."

PIERRE CHARPIN

who is he?

This sculptor's son, born in Saint-Mandé near Paris, graduated from the school of fine arts in Bourges. He had a revelation at the end of the 1970s while flicking through the design review *Domus* in which he discovered Italian avantgarde design movements such as Studio Alchimia and Memphis. He showed his works for the first time at the Galerie Nestor Perkal in Paris, and then, during an assiduous visit to the Milan furniture fair, he caught the attention of important manufacturers such as Zanorra and Brunati, sensitive to the precision of his elegant style.

his style

Simplicity, rigor, and harmony are the terms that best define Pierre Charpin's creations. Whether with a set of glasses, a couch, or a vase, this measured designer enhances the plastic beauty of pure form. He borrows the rejection of excessive detail and a sense of proportion from the masters of American minimalist art such as Donald Judd, Carl André, and Robert Morris. But sobriety does not mean severity. This lover of beautiful materials also enjoys experimenting with ancestral skills, as he did with pottery in Vallauris, a French village renowned for ceramic art, or with glass at the CIRVA (International Center for Research in Glass and the Fine Arts) in Marseille. A subtle colorist, he knows how to give his productions a soft, sensual feel.

best of

1997 Laminated table with tinted wooden ridges, Post-Design ❖ **1998** Slice armchair, Brunati ❖ **2002** Stands, Design Gallery, Milan ❖ **2004-05** Eau de Paris decanter ❖ Elected designer of the year at the Salon du Meuble, Paris

Slice, produced by Brunati, distributed by Kreo, 1998.

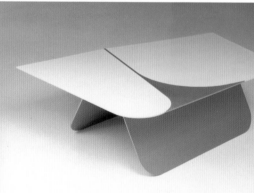

Lacquered and polished aluminum coffee table, Platform exhibition, Galerie Kreo, 2006.

Ceramica 2, ceramic vase, Oggetti Lenti, Designgallery, Galerie Haute Définition, 2005.

He said:
"I give more importance to the impact my objects have on the environment that contains them, than to their actual shape."

MATALI CRASSET

who is she?

This farmer's daughter was born in Châlons-en-Champagne in France in 1965. She studied marketing before taking a degree at the ENSCI (National School of Industrial Design) in Paris in 1991. She earned her spurs with Denis Santachiara, then with Philippe Starck who, in 1993, employed her as part of the Thomson Multimedia team of which she was later in charge, before opening her own set-up in 1998.

her style

Matali Crasset's designs tell us about our age: its minimalist lines revive the contemporary customer's need for essential shapes, overwhelmed as he is by useless products, while responding to a fascination for new technology. She likes open spaces, adjustable furniture, convivial objects, and flashy colors. The Hi Hotel in Nice, which she planned in its entirety, totally rethinks our ways of living. Indeed, Matali Crasset does not defend a style but rather a state of mind: her extra bed entitled Quand Jim monte à Paris (When Jim Comes to Paris), her stool that transforms into a mattress for an afternoon nap labeled Téo de 2 à 3 (Téo from 2 to 3), or her jigsaw puzzle couch Permis de Construire (Building Permit) offer us everyday scenarios. Her vibrant, rigorous, and poetic environments adapt to all circumstances and breathe an air of twenty-first century optimism into our homes.

best of

1995 Quand Jim monte à Paris, hospitality column ✸ **2000** Permis de Construire, a couch that can be dismantled like a child's game ✸ **2004** Decompression Space armchair ✸ **2005** Opening of boutique Lieu Commun (Common Space) with the fashion designer Ron Orb and the music label Fcommunication ✸ **2006** Elected designer of the year by the Salon du Meuble, Paris

Permis de Construire, Domeau & Pérès, 2000.

Nextome, chair/magazine rack, Domeau & Pérès, 1999.

Rendez-vous, bedroom at Hi Hotel, Nice, 2003.

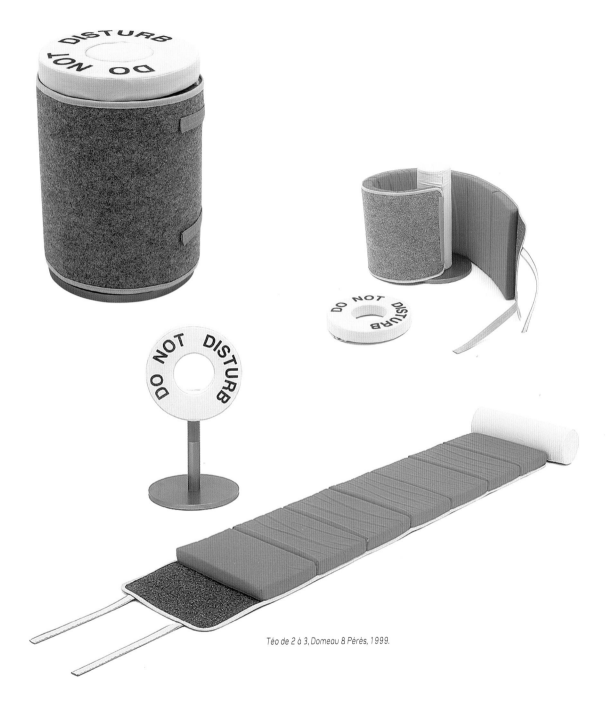

Téo de 2 à 3, Domeau & Pérès, 1999.

She said:
"Fortunately, I can't imagine all the things people will do with my creations!"

DROOG DESIGN

who are they?

Droog Design (pronounced "drog") means "dry design." This experimental Dutch laboratory of ideas, which is nevertheless amenable to the world of commerce and industrial production, was founded in 1993 by the designer Gijs Bakker and the design historian Renny Ramakers. Their role? To find talented individuals from all fields (design, graphic design, fashion design, architecture) whom they entrust with projects in order to establish the Droog Design collection on the basis of a shared ideology.

their style

More than a style, Droog Design represents a biting spirit which can be summed up in a few words: humor, provocation, creativity, and recycling. Add to this an overflowing imagination and you get a good measure of poetry. The protagonists of this label fight against waste through the reappropriation of existing elements to which they give a new unreserved use: for example, old drawers stacked anyhow are transformed into a chest of drawers, and a dozen milk bottles metamorphose into a lamp. Several big names in contemporary design—Jurgen Bey, Marti Guixé, Richard Hutten, Marcel Wanders, Hella Jongerius, Rody Graumans, Joost Grootens—started out or have collaborated with this exceptional, experimental platform that is not far from the concept of the ready-mades invented by Marcel Duchamp in the early twentieth century. Today, this group accounts for more than two hundred objects created by more than a hundred designers.

best of

1991 Chest of Drawers, Tejo Remy ⬦ **1993** 85 lamps ceiling light, Rody Graumans ⬦ **1994** Bench the Cross, Richard Hutten ⬦ **2001** Chest Box, Jan Konings ⬦ **2002** Tree Trunk bench, Jurgen Bey

Urn vase, Hella Jongerius, 1993.

Milk Bottle lamp,
Tejo Remy, 1991.

Chest Box, Jan Koonings, 2001.

Chest of Drawers, chest of drawers made out of old drawers held together by a belt, Tejo Remy, 1991.

They said:
"Droog Design aspires to be a collection of cultural fragments which are gathered, reorganized, and redefined."

NAOTO FUKASAVVA

who is he?

Born in 1956 in Yamanashi, Japan, Naoto Fukasavva graduated in industrial design from Tama Art University in 1980 and has since led a flawless career. He began work at Seiko-Epson just after graduation and stayed there for eight years. In 1989, he moved to the United States and joined the Idéo company. Back in Japan in 1996, he created Idéo Japan. In 2003, he opened his own studio, Naoto Fukasavva Design in Tokyo. He became the artistic director of the ±0 brand which makes objects for the home on a large scale and joined Muji's board of directors.

his style

Naoto Fukasavva should be placed in the category of those who do not seek, but find. "Design and creation," he says, "are not a result that we achieve but one that is already there." Observing our behavior, detecting unexplored areas, understanding what we need to feel good, responding with a touch of humor: such is this man's way of working, patient and undeterred by changes in fashion. Rigor and simplicity do not preclude empathy. This is true of his famous wall-mounted CD player or his wireless eight-inch television. As soon as they are installed, they are already a part of the family.

best of

1999 CD player, Muji ❖ **2003** 8-inch LCD television, ±0 ❖ **2005** Air humidifier, ±0 ❖ **2006** Muku collection of tables and chairs, Driade Aleph

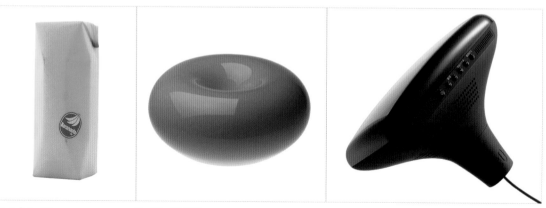

Haptic/Juice Skin,
Takeo Paper Show, 2004.

Humidifer, ±0, 2005.

Eight-Inch LCD TV, ±0, 2003.

CD player, Muji, 1999.

He said:

"The less an idea needs explaining, the stronger it is. The process is never simple, but our aim is to ensure that the result always is."

CHRISTIAN GHION

who is he?

For a long time, Christian Ghion sought his way. With a law degree in his pocket, he attempted, unsuccessfully, the entrance exam for the prestigious French École des Sciences Politiques school in Paris and the FEMIS (the French National School of Image and Sound). In the end, he turned to what he had always liked doing: drawing. He took the ECM degree (French degree in furniture design) in Charenton and, in 1987, opened a design studio with the architect Patrick Nadeau, a collaboration which was to last ten years. After a friendly separation, he now works alone.

his style

Christian Ghion's beautifully structured creations are both elegant and sensual. Their colors highlight their streamlined forms, while spreading a positive energy in the image of their creator's dynamism. Funny, loudmouthed, and a bringer-together of talents, Ghion willingly plays the band leader. His boxer's physique contrasts with his obsessive taste for refined finishing touches, and soft and feminine materials. He calls himself "the gypsy of design" and lugs around a few beautiful successes in his caravan: the famous Butterfly armchair and the Shadow chaise longue both took shape beneath his master pencil stroke.

best of

2000 Creation of the "Design Lab" section for the Salon du Meuble in Paris ❖ **2000** Rug for Tarkett Sommer ❖ **2002** Shadow chaise longue, Cappellini ❖ **2004** Butterfly Kiss chair, Sawaya & Moroni ❖ **2004** Chantal Thomass boutique, Paris ❖ **2005** Jean-Charles de Castelbajac boutique, Paris

Butterfly Kiss chairs, Sawaya & Moroni, Milan, 2004.

Mise en Plis table, Tendo, Japon, 1996.

He said:
"The sensual character is a willed part of identity. And I took a real pleasure in assuming things."

KONSTANTIN GRCIC

who is he?

Born in Munich in 1965, he first studied at a private professional school, Parnham College, Dorset, and then at the Royal College of Art in London. There, he met Jasper Morrison with whom he teamed up for a while. In 1991, he opened his own agency in Munich. Considered even from his student days as one of the most talented designers of his generation, he saw his commissions book fill up almost instantly. He has collaborated since then with numerous international companies such as Whirlpool, Driade, Flos, Cappellini, and Muji.

his style

From what planet do Konstantin Grcic's creations come down? They are very different from other models in circulation and have a lunar feel to them. Whether a micro-kitchen for Whirlpool which could be taken for a flying saucer, or a hybrid half-stool–half-table, or a bookcase with lateral sides made of sticks resembling spears, their appearance is disconcerting and requires a moment of adjustment. Although Konstatin Grcic defends modest, scientific design. Without complications. When faced with this capacity to reinvent the most basic of typologies, we find him intelligent and poetic.

best of

1998 Mayday lamp, Flos ❖ **2000** Whirlpool micro-kitchen ❖ **2001** Chaos chair, ClassiCon ❖ **2004** Design of the exhibition *Design en Stock*, 2000, objects from the Fonds National d'Art Contemporain in Paris ❖ **2005** Diana table, ClassiCon ❖ **2006** Mars chair developed as a template.

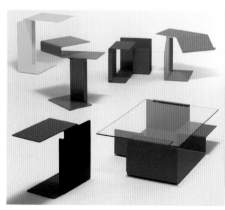

Diana Family, ClassiCon, 2002.

Go, Authentics, 1999.

Chef Plan, Whirlpool, 2000.

He said:
"For me, physical comfort arises from intellectual comfort."

ORA-ÏTO

who is he?

Born in Paris in 1977, Ora-ïto studied at the ESDI (School of Industrial Design) in Paris and worked as an intern in various architectural agencies. In a hurry to make his reputation however, he had the idea of creating virtual digital images of designer label products without the slightest authorization. For example, a back-pack for Louis Vuitton, a watch for Nike, and other utopian forms for Bic and Apple. He came close to several court cases, but in the end the brands got something out of the media coverage. And Ora-ïto made his name overnight. Endowed with a surprising gift of the gab, the young designer now works on all design fronts with an innate sense of concept, a lot of clairvoyance, and a crazy desire to leave his mark.

his style

The name of Ora-ïto—a pseudonym invented by its author from his own surname for its Eastern sound—perfectly fits his purist style of soft curves. After only a few years of practice, the productions of this design meteor show a vaguely futuristic fluidity, but create a benevolent relationship with their user. Refusing to specialize in one domain or another, Ora-ïto is sometimes a graphic designer, sometimes an interior designer, sometimes an Internet site designer, and sometimes an adman. Or all at once. In the end, whether he designs a perfume bottle, lamp, or poster, his works have a quality in common: they are all in touch with or precede their time. And so the Heineken drinks can transformed into a bottle or the Enterprise ceiling light flirt with the science fiction and comic strips which clearly feed his imagination.

best of

1998 Pirating of designer labels into digital images ✷ **2002** Interior design of the CAB nightclub in Paris ✷ **2003** Petal armchair, Cappellini ✷ **2004** One Line lights, Artémide ✷ **2004** Toyota showroom ✷ **2004** Nike House, Paris

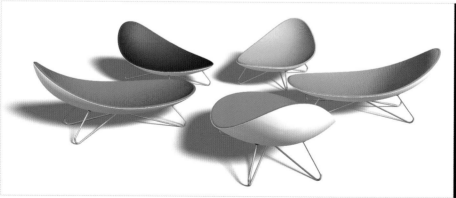

Compoverte, Cappellini, 2002.

*Back Up,
Vuitton/Jalouse, 1999.*

Luxury Project, Heineken, 2002.

He said:
"A fragrance or a car, it's the same thing."

HELLA JONGERIUS

who is she?

She was born in 1963 at De Meer, in the Netherlands. In 1993, with a degree from the Eindhoven Academy for Industrial Design in her pocket, she joined Droog Design, a collective specialized in recycling and reuse that brings together young designers. She opened her own studio, the JongeriusLab, in Rotterdam in 2000. Her experimental taste resulted in the creation of hybrid objects and her work has rapidly gained a reputation. She is regularly invited across the world to present her productions and collaborate with various large industrial firms.

her style

A non-conformist attitude, a capacity to bring together opposing materials, and a natural sense of humor feed the work of this female Professor Calculus (a character from the comic strip *Tintin* by the Belgian author Hergé). She enjoys nothing more than knocking together unnatural associations, testing improbable comparisons, and surprising her public by creating unexpected links between yesterday and today. Witness, for example, the embroidered china plates that reinterpret traditional Delft designs, or the synthetic chair inspired by African prayer stools. With one foot in the future, her Bed in Business bed is endowed with the latest technologies. She is also gifted in industrial design: her collaborations with Maharam textiles or with the Porzellan Manufaktur Nymphenburg lived up to the expectations of both commissioners and customers.

best of

1998 7 pots / 3 centuries / 2 materials ceramic vase ❖ **2000** Long Neck 8 Groove bottles ❖ **2002** Blizzard bulb ❖ **2003** Collection of furniture fabrics designed for Maharam and presented at Moss, New York ❖ **2004** Elected designer of the year by the Salon du Meuble, Paris ❖ **2005** Series of vases, PS collection, Ikea

Blizzard bulb,
limited edition, 2002.

7 pots / 3 centuries / 2 materials, 1998.

Embroidered Tablecloth, linen, cotton, porcelain, limited edition, 2000.
Specially created for the Keramiekmuseum Princesshof, Leewarden, The Netherlands, distribution JongeriusLab, Rotterdam.

She said:
"People are fed up with change and expect objects that mean something."

MATHIEU LEHANNEUR

who is he?

Born in Rochefort in 1974, Mathieu Lehanneur graduated from the ENSCI school of industrial creation in Paris in 2001. In 2004, his first show was at the contemporary arts center in Brétigny in France and a year later the Museum of Modern Art in New York purchased ten Objets Thérapeutiques (Therapeutic Objects) exhibited in *Safe Design Takes on Risk*. In 2006, he was awarded the Paris Grand Prix. The exhibition of Elements for the VIA, at the Salon du Meuble, brought him to the attention of an enthusiastic public.

his style

Passionate about exploring design, Lehanneur creates objects imbued with fantasy and poetry for tomorrow and beyond that he hopes will transform our behavior. He enjoys imagining scenarios and creating likable objects or attractive environments capable of improving our everyday lives. For his work, he consults at length with specialists (geneticists, doctors, biologists, etc.), and undertakes a huge amount of documentary research. Then, he draws. There is nothing approximate about the string of bead-like doses of medication of Medicine by the Centimeter, his house for stray cats, or his strange devices for purifying the air, muffling disturbing noises, and adjusting temperature. Lehanneur attempts to reconcile reality and dream.

best of

2001 Ten Objets Thérapeutiques ✺ **2005** A sign project conceived in the form of the morning star for the municipal library and theater in Brétigny-sur-Orge ✺ **2006** Refurbishment project for a new restaurant chain, Fast Quality Food ✺ **2006** Elements for the VIA at the Salon du Meuble, Paris

Gary Copper, radiator with copper ribs, 2005.   *Le Feutre Thérapeutique, Objets Thérapeutiques series, 2001.*

dB, Elements, 2006.

He said:
"How far can the designer flirt with architecture and, conversely, how far can he interact with and act upon the individual?"

INGO MAURER

who is he?

Born in 1932 in Germany on the island of Reichenau, Ingo Maurer studied typography and graphic design in Switzerland. He then set up in the United States where he worked as a freelance designer. Back in Munich, he founded the production company Design M in 1966, and made his reputation with the Bulb desk lamp. He specialized in light, reinventing lighting with strange lamps and original luminous installations for public spaces and exhibitions.

his style

Ingo Maurer makes light sparkle, shadows dance, and reflections shimmer. His works fill us with wonder. Spectacular or humble, they remind us of the enchanted world of childhood and naive tinkering. Like a magician, this designer of the intangible creates a floating universe through a discreet use of the latest technologies. His poetic creations transport us to a land of dreams and bewildering tales. This is true of the humorous installation entitled Wo bist du, Edison, jetzt, wo wir dich brauchen? (Where are you Edison when we need you?), or his delightful little lamp Lucellino in the form of a simple light bulb fitted with two goose-feather wings, like an angel, ready to fly away.

best of

1966 Bulb desk lamp ✻ **1980** Lampampe paper lights ✻ **1983** Ilios halogen light ✻ **1992** Lucellino lamp ✻ **1997** Wo bist du, Edison, jetzt, wo wir dich brauchen? ✻ **2003** Led Table light table

YaYaho, 1984.

Golden Ribbon, 1997.

Lucellino, 1992.

He said:
"My wish is to make modest things—objects that get under your skin little by little."

ALESSANDRO MENDINI

who is he?

Born in Milan in 1931, this well-known figure of postmodern Italian design obtained a PhD degree in architecture from Milan Polytechnic University in 1959. A visionary theoretician full of ideas, totally against any industrial pressure, in 1973 he co-founded the Global Tools design school, the forerunner of Studio Alchimia which was to break with international modernist tradition in the early 1980s. He invented the concepts of re-design and banal design by reusing modern and contemporary furniture classics. In 1989, he opened the Mendini studio with his brother. He was also artistic director at Swatch and a consultant for Alessi. The editor of several magazines such as *Casabella*, *Modo*, and *Domus*, his pertinent analyses had an important influence on the current generation of designers and propelled Italian design into the spotlight where it would remain for a long while.

his style

Eclectic but recognizable. Provocative, he professes to work with "only with fifty or so norms indexed in the computer, which are recycled and re-combined each time." In fact, Mendini likes to dramatize the everyday, staging common objects and repainting domestic icons. Color takes center stage, giving new life to chairs, coffee pots, and kitchen utensils. With no time for mass-produced characterless objects, he contrives to reanimate the indispensable and the incontrovertible. This flamboyant designer has succeeded in giving design both a decorative and a critical dimension.

best of

1978 Siège d'Auteur: Wassily de Marcel Breuer ❖ **1978** Poltrona di Proust armchair, Studio Alchimia ❖ **1985** Decoration of the Renault Super 5 with Studio Alchimia ❖ **1992** Panton Proust ❖ **1994** Anna G corkscrew, Alessi

Anna G, Alessi, 1994.

Panton Proust chair, series of two prototypes, Kreo Gallery, 1990.

Mendini Studio, bus stop, Hanover, Germany, 1994.

Siège d'Auteur: Wassily de Marcel Breuer, redesign of seats from the modernist movement, Mendini Studio, 1978–83.

He said:
"Each object is a symbol, is a fetish, is significant; is the fruit of utopias, moods, and loves."

JASPER MORRISON

Born in London in 1959, this star of design studied at Kingston Polytechnic in Surrey, then at the Royal College of Art in London, before taking advantage of a scholarship to go to Berlin where he frequented the Hochshule für Kunst. With this rich experience under his belt, he opened a studio in London in 1986, then another in Paris in 2002. Rapidly recognized for his talent, he quickly garnered international fame. In the late 1980s, his zen furniture, in complete contrast to the baroque style then in fashion, was a breath of fresh air in an atmosphere saturated by chaotic productions.

his style

For this rigorous designer, everything relies upon the purity of drawing. Has Morrison brought functionalism back into fashion? Yes, but in a more flexible fashion than that defended by the ideologists in the early twentieth century. With him, purity means fluidity, delicacy means lightness, and simplicity means discretion. This British designer's practice, the darling of important design manufacturers such as Cappellini and Vitra due to his numerous commercial successes, is equally adored by a public for whom beauty rhymes with discretion. And his sense of appropriateness is directed toward everyday objects which he manages to make desirable: his kettle and coffee machine for Rowenta and his cutlery for Alessi have become classics of domestic design.

best of

1987 Designs the Reuter's News Center for the *Dokumenta VIII* exhibition in Kassel ❖ **1998** Plywood chair ❖ **1999** Low Pad armchair and Hi-Pad chair, Cappellini ❖ **1999** Glo Ball lamp, Flos ❖ **1999** Air chair, Magis ❖ **2000** Fits out the Tate Modern in London ❖ **2002** ATM desk, Vitra ❖ **2004** Rowenta kettle ❖ **2004** Soft Sim Low chair, Vitra

Low Pad chair, Cappellini, 1999.

Hanover Streetcar, Üstra, 1997.

Glo Ball, Flos, 1999.

He said:
"Too much personal expression can harm group expression."

MARC NEWSON

who is he?

An emblematic designer of the 1990s, he was listed by *Time* magazine as one of the hundred people who shape our lives. Handsome, rock 'n' roll at heart, and a close friend of many a top fashion model, this Australian, born in Sydney in 1963, studied jewelry and sculpture at the Sydney College of Arts. He never misses an occasion for publicity. Crazy about surfing, he knows how to catch the wave and make radical moves. He earned his spurs at the VIA and opened his first office in Paris in 1991, before moving to London in 1997. He continues, however, to move back and forth between the two capitals. Having signed a few stunning pieces with aerodynamic contours, he turned to mass marketing in order to reach a wider public.

his style

Nothing scares Marc Newson, especially not new manufacturing methods. His tastes run toward carbon fiber, composite materials, Corian, etc., but he also uses cane, string, and cardboard. Probably an athlete's reflex, his drawings privilege organic forms, ellipses, and movement. He designs a concept-car, athletic wear, a bicycle, the interior of an airplane, or a hotel with equal ease. He is delighted with his collaborations with Tefal or Nike because, he says, "I would like my mother to be able to go into a shop in Australia and buy something I designed." Pragmatic but above all an artist, he has already designed several icons of contemporary furniture, including the Wicker chair and the Alufelt chair.

best of

1986 Lockheed Lounge chaise longue, Pod ❖ **1988** Wood Chair, Cappellini ❖ **1990** Wicker Chair chaise longue, Idee ❖ **1993** Chosen designer of the year at the Salon du Meuble in Paris ❖ **1993** Alufelt chair, Pod ❖ **1998** Dassault Falcon 900B ❖ **1999** 021C concept-car, Ford Motor Company ❖ **2003** Designed Qantas business class ❖ **2003** Series of saucepans for Tefal ❖ **2004** Zvezdochka shoes, Nike

Wood chair, Cappellini, 1988.   Hemipode Gold, Ikepod Watch Co., 2002.   Falcon 900B Dassault, 1998.

Azzedine Alaïa boutique, Paris, 2006.

He said:
"As with Samsonite, Ideal Standard, Tefal, and Ford, people who buy my products don't necessarily know that I am behind them."

PIERRE PAULIN

who is he?

This figure of French design has been internationally appreciated since the 1960s. Born in Paris in 1927, Pierre Paulin studied at the Camondo school. His tastes led him toward organic forms such as those practiced by Aalvar Alto, and Charles and Ray Eames. After the Second World War, he worked with the Thonet brand, for which he designed in particular a wooden desk, then, in 1959, he began a rich collaboration that the Dutch production company Artifort for which he created models that are references today. In 1972, he decorated the apartments of President Georges Pompidou of France at the Élysée and François Mitterand's office in 1984. In 1995, he moved to the Cévennes region of France, and while he considerably reduced his activity, his reputation began to grow. He died in 2009.

his style

The design of Pierre Paulin's seats is almost perfect. His furniture features simple shapes made from a rigid structure padded with foam and covered in colored jersey fabric. Their sensual and fluid outlines combine harmony and daring. The perfectly proportioned Tongue chair offers its voluptuous forms to lie on, while his Ribbon chair makes for a stately but comfortable seat thanks to a knowing combination of full and empty forms. They have become icons of the 1960s, yet Paulin's furniture is ageless.

best of

1959 Mushroom chair, Artifort ✿ **1964** Tongue chair, Artifort ✿ **1966** Ribbon chair, Artifort ✿ **1970** Amphis couch, Artifort ✿ **2004** Exhibition at the Meuron Gallery

Amphys couch, Alpha International, Casson Méron Gallery, 1970.

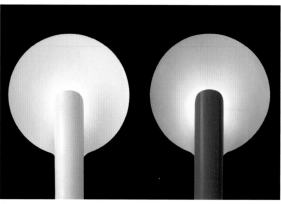

Élysée lamp, Verre et Lumière, Casson Méron Gallery, 1972.

Fauteuil 582, known as Ribbon, Jack Lenor Larsen fabric, Artifort, Casson Méron Gallery, 1966.

He said:
"Furniture must first be useful to the person who uses it and comfortable.
If there is some poetry, then all the better ... but it comes after the rest."

GAETANO PESCE

who is he?

Born in La Spezia, Italy, in 1938, Gaetano Pesce graduated from the architecture faculty of the University of Venice. In the early 1960s, he became a freelance interior decorator and worked closely with proponents of kinetic art and arte povera. Simultaneously, he worked on furniture and interior design. He moved to Paris in the 1970s, then from 1982, he chose to live in New York. He feels restricted by the title of designer, and prefers to call himself "a pre-visionary creator."

his style

Gaetano Pesce's productions stand out by their formal originality and their expressiveness. Constantly fighting against the formatting of functional design, he plays with materials, in particular plastics, like a sculptor. He likes the unfinished and the irregular. He has attempted more than once to subvert manufacturing processes in order to obtain objects that were unique, despite the fact that they had been industrially made. Sensuality, humor, and disorder are the focal points of his research. The Up series of inflatable seats and the Dalila chairs testify to his taste for organic and anthropomorphic forms inspired by the female body.

best of

1969 Series of Up chairs ❖ **1970** Moloch lamp ❖ **1970** Up 7, Il Piede ❖ **1980** Dalila chair ❖ **1980** Tramonto à New York, Cassina ❖ **1987** I Feltri chair, Cassina ❖ **1987** Sansone I table ❖ **2003** Nobody's Perfect furniture series ❖ **2004** Alda lamp (named after his mother)

Sunset in New York, Cassina, 1980.

Carenza bookshelves, 1993.

Alda lamp, 2004.

He said:
"The human being is half-way between instinct and reason.
The Alda lamp is meant to make a small contribution to this debate."

OLIVIER PEYRICOT

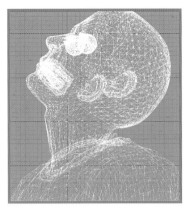

who is he ?

Born in 1970, Olivier Peyricot graduated from the ESDI school of industrial design in Paris. He co-founded the IDSland agency with Kiki Chanchus. His work was shown for the first time at the Neotu Gallery, in 1995, with a collection of furniture pieces called Stratégies Individuelles designed to offer "new modes of living." He made several objects over the years for Laurence Dumaine's Galerie Axis, as well as bags for saddler Éric Laville. Particularly interested in expressing new contemporary behavior, Peyricot threw himself into creating futuristic design with innovative scenographies. In 1999, he designed the exhibitions Hypothèse de Collection and En Quête d'Objets at the Centre Pompidou. From 1998 to 2001, he edited the magazine Mobile.

his style

The notion of style is totally foreign to this designer, who conceives of his practice as a succession of experiments that result in the establishment of a constructive interactivity with the user. He believes it is urgent to find innovative solutions to the contemporary problems of nomadism, precariousness, and individualism. His fondness for progressive design discards a static, repetitive conception of contemporary furniture, preferring adaptable and user-friendly objects. He suggests, for example, that we adopt his Body Props, light and discreet ergonomic human props that adapt to the body's various positions.

best of

1998–2001 Editor in chief of Mobile ✷ **1999** Hypothèses de Collection, Expérience Glassex ✷ **1999** Exhibition design of En Quête d'Objets, Centre Pompidou ✷ **2000** Club Med 2000, furniture and design for holiday village ✷ **2002** Body Props, Edra

Réaction bookcase,
IDS France, 1997.

Vivre Nu trays, for Corian, exhibition at the Centre Culturel Français in Milan, Italy, April 2006.

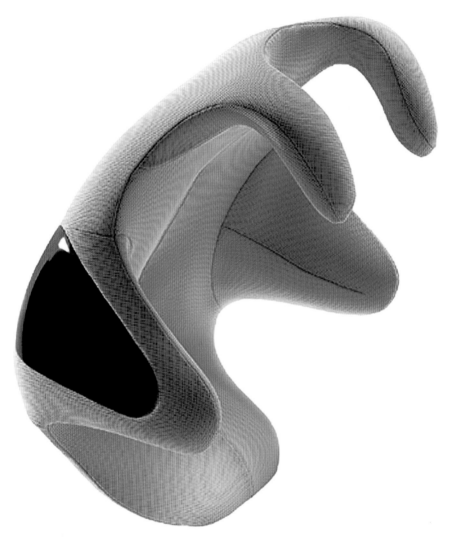

Wear Your Seat, VIA, 2003.

He said:
"I choose to design useful things, furniture that is "in anticipation of,"
ready to reply to an individual user's plans."

CHRISTOPHE PILLET

who is he?

Born in France in 1955, with a degree from the school of decorative arts in Nice in 1985, he received a Masters degree from the Domus Academy in Milan. He plugged right into the contemporary design concerns of the time thanks to his collaboration with Martine Bedin, formerly of the Italian group Memphis. Back in Paris in 1988, he joined Philippe Starck's team and worked with him for five years. In 1991, the XO firm produced his first furniture pieces. He then opened his own studio in 1993 where he worked freelance and collaborated with various furniture producers, Cappellli, Artelano, Magis, and Driade, as well as companies such as JCDecaux, Daum, Lacoste, and Axis. He has designed many shops and hotels.

his style

For Christophe Pillet, the very notion of style is obsolete. While conscientious about fulfilling commissions, he brings to each project that little extra which makes all the difference. His own way of giving shape to the unexpressed. Familiar with experimental procedures, he rejects excess and keeps his distance. Seen as the leader of the "French touch" generation, he is the author of sober, elegant, and perfectly finished works. This lover of life, who was artistic director of the Paris Salon du Meuble's Metropole space, knows how to combine modernity with comfort, as well as how to impose his works without bravado—works that are recognizable by their reassuring presence.

best of

1998 Y's chair, Capellini ❖ **2000** Rodolphe Menudier boutique, Paris ❖ **2002** Lobby sofa, Domeau & Pérès ❖ **2004** Artistic director of the Metropole space of the Salon du Meuble in Paris ❖ **2004** Catherine Malandrino boutique, New York ❖ **2005** Hotel Sezz, Paris

Agatha Dream chaise longe, Ceccotti, 1999.

Sunset armchair, Cappellini, 1998.

Air Can lamp, Mazzega, 1999.

He said:
"Today, we live differently, we talk differently, we communicate differently, and design participates in this by accompanying these developments."

ANDRÉE PUTMAN

Andrée Putman's career has taken a series of short cuts. Born in Paris in 1925 to a family of bourgeois intellectuals, she intended to become a pianist. But her passion for the art and artists of her time led to a change in direction. She became an art and decoration jounalist for *Elle* magazine, then for *Marie-Claire*, and in 1968 founded Mafia, the first French design agency. Her main client was the Prisunic supermarket chain. She also founded the company Créateurs et Industriels with Didier Grumbach. Ten years later, she opened another agency, Ecart, and embarked upon the reproduction of forgotten furniture by great modern designers: Eileen Gray, Pierre Chareau, Robert Mallet-Stevens, and Fortuny. In 1984 the refurbishment of the Morgans Hotel in New York launched her international career.

her style

Andrée Putman launched the French bourgeois bohemian loft style. She was an early champion of the association of old and new, the cheap and the precious, and new twists. An alchemy that she masters to perfection. It was she who rescued the famous Parisian chic from its floundering in backward-looking narcissism. She lays claim to an eclectic taste, the use of black and white, and lines that never go out of fashion. She is thrilled to be able to refurbish museums: the CAPC museum in Bordeaux, France, and the lobby of the Guggenheim Museum in New York. As for private commissions, she likes to design modern, pleasant, human, dreamlike spaces. She has just created a fragrance exclusive to the Colette boutique in Paris. The ultimate consecration!

best of

1978 The reproduction of modern classics by Ecart ❖ **1984** Morgans Hotel, New York ❖ **1990** CAPC museum, Bordeaux ❖ **1991** Interior decoration of Concorde for Air France ❖ **2001** Ritz Carlton Hotel, Wolfsburg ❖ **2004** Pershing Hall Hotel, Paris ❖ **2005** Préparation Parfumée (Perfumed Preparation)

French Ministry
of Education, 2002.

Préparation Parfumée,
fragrance bottle, 2005.

Hotel Sheraton-CDG, Charles De Gaulle Airport, Paris-Roissy, 1995.

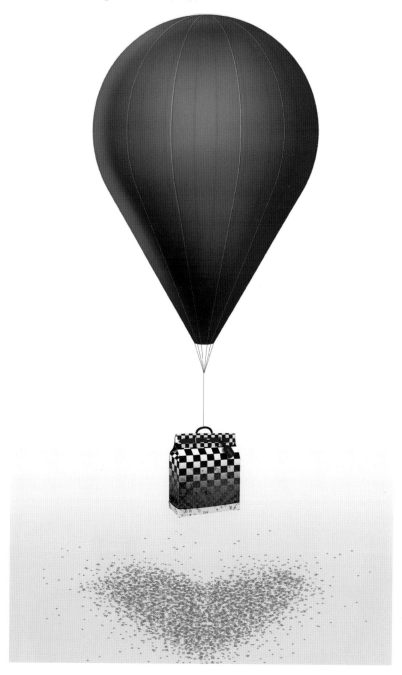

Streamer, Louis Vuitton-Andrée Putman, 2006.

She said:
"I have this great obsession with the truth. Making something true, something experienced, is nothing like a trick or pretence."

RADI DESIGNERS

who are they?

Founded in Paris in 1992, having finished their degrees at the ENSCI/Les Ateliers in Paris, this fluid European collective started life with five members: Laurent Massalou, Olivier Sidet, Robert Stadler, Florence Doléac, and Claudio Colucci. Then there were three designers, then four, then back to three again. Doléac and Colucci work independently today. In 1998, the whole group gave its first exhibition at the Galerie Emmanuel Perrotin, at the invitation of Didier Krzentowski and, in 1999, the Fondation Cartier pour l'Art Contemporain entrusted them with an installation. This was shown the following year at the Park Hyatt hotel in Tokyo and at the Insa Art Center in Seoul. In 2001, the Sandra Gering Gallery in New York organized their first exhibition in America. In Paris, Galerie Kreo and Galerie Tools distribute limited editions of their products.

their style

Confusing the senses and championing reuse, the off-beat style of these young designers flirts with contemporary art. Skillfully pushing at the limits that separate creativity and function, they invest different fields of design with total freedom and a unique capacity for telling stories. With them, a bench takes the form of a dog, a cookie placed on the edge of a cup resembles a splash of coffee, and a public water fountain has the abstract profile of a little water carrier. Involved for a long time with prototypes and art galleries, they were able to diversify their activity and respond to mass marketing with simplicity and humanity. For example, the attractive range of electrical appliances for Moulinex and the Marlboro matchbox.

best of

1998 A range of measures for Ricard ⚬ **2000** Glasses for Schweppes ⚬ **2001** Le Feu d'Issey Light cosmetic bottles ⚬ **2003** Matchboxes for Marlboro ⚬ **2005** The "Principio" range of electrical appliances for Moulinex

Meal-tray, Air France, 2000.

Transparent Cake, tableware, In'nova, 2003, Tools Galerie, 2005.

Lampshade for a candle, prototype, CRAFT, 1997.

50 DESIGNERS and their signature works

Coffee Drop Splash, prototype, 1994.

They said:
"We like to juggle with the obvious things in life, the gestures and the typologies of objects-products-furniture-gadgets."

KARIM RASHID

who is he?

Born in Cairo, Karim Rashid is half-Egyptian and half-English and was brought up in the United Kingdom and in Canada. He graduated with a degree in industrial design from the University of Ottawa in 1982 and then continued his studies with a year in Milan under the aegis of Ettore Sottsass. Back in Canada, he worked for seven years for KAN Industrial Designers and in 1985 co-founded Babel Inc. and North Studio. In 1993, he opened Karim Rashid Inc. in New York. He responds to demands in fields as varied as wallpaper, cutlery, fashion, 3D graphic design, the design of computer programs, jewelry, lamps, and furniture, while continuing his personal work as an artist.

his style

Colorful and eclectic, Karim Rashid's style does not fit into any precise, formal category. Occasionally flexible, sometimes rigorous, his lines mix many and varied influences. A similar energy runs through all of them however. Generous curves emphasize their theatrical appearance. Prolific, intuitive, and dynamic, Karim Rashid draws quickly. The number of his productions is impressive: it is well over a thousand. He has worked for Prada, Issey Miyake, and Alessi, among others. Although passionately interested in new technologies and materials, he also believes that it is time to reduce over-consumption. The flow of his projects does not seem to be lessening however.

best of

2001 Sunlite seat ❖ **2002** Superblob couch, Edra ❖ **2003** Kurve chair armchair, Nienkamper ❖ **2005** Ya Table table

Melissa Ananha, Melissa High, 2005.

Garbo Can, Umbra, 1996.

Blue Skwarim, LaCie, 2006.

Mystik chair, prototype for the Milan furniture fair, Domus, 2002.

He said:
"I want to change the world." (The title of his book, published by Taschen.)

ÉRIC SCHMITT

who is he?

Born in 1955, Éric Schmitt originally turned toward experimental music and sculpture before showing furniture for the first time in 1987 at the VIA in Paris. He then joined the neo-baroque group of designers Les Nouveaux Barbares who, like him, tackled techniques of wrought iron, casting, metal, and stone. There he mixed with Olivier Gagnère, Élisabeth Garouste, Mattia Bonetti, and Jean-Claude Gleize. He then entered the Neotu Gallery. In 1996, this experienced designer set up a studio on the edge of the Fontainebleau forest, and began a long collaboration with Christian Liaigre, designing bronze lamps and coffee tables. His own work consists of furniture, lamps, stair grids, and ramps for often prestigious public and private places. He regularly shows his original creations in galleries across Europe and the United States.

his style

Éric Schmitt is probably one of the most creative inheritors of the French art deco tradition. He is part of the generation of designers who, in the 1980s, sought to update ancestral skills in a contemporary spirit. He cultivates an ambiguous practice situated on the borderline between sculpture, architecture, and design. A defender of traditional materials, he reinvents their impact: in this way, industrial alloys take on a precious appearance and refined materials a rough aspect. Original forms, clean lines, and discreet ornament give each of these pieces a surprising presence.

best of

1989 Boule bronze lamp, En Attendant les Barbares ❖ **1991** Stairway, lamps, and objects in wrought iron and bronze for the Hôtel Montalembert in Paris ❖ **1993** Lampe à Barreaux, Galerie Neotu ❖ **1999** Tabou table in dark bronze, Christian Liaigre ❖ **2006** Opposite desk in two-toned bronze, Ralph Pucci, New York

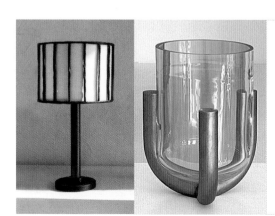

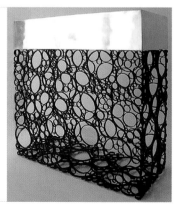

Lampe à Barreaux,
Galerie Neotu, 1993.

Round Cross vase,
at Ralph Pucci, 2006.

Taboo stool, Christian Liaigre,
1999.

Lace console table,
Éric Schmitt studio, 2004.

30 DESIGNERS and their signature works

Galet console table, Mobilier National, 2005.

He said:
"My work is the consequence of the sculptor who I almost am, of the architect I would have liked to have been, and the designer who I am not completely."

ETTORE SOTTSASS

who is he?
The high priest of Italian design, Ettore Sottsass was born in Innsbruck in 1917 to an architect father who had studied in Vienna with Otto Wagner. After graduating from the Turin Polytechnic in 1939, he worked at first with his family, then opened his first agency in 1947. In 1958, he became a consultant for Olivetti, for whom he designed numerous emblematic objects of 1960s Italian industrial design, including the famous red Valentine typewriter which was shown at the MoMA in New York in 1969. Ten years later, trips to India and the United States having worn away his faith in functional and rational forms, he joined the Studio Alchimia before founding the Memphis group in 1980, with other designers (including Martine Bedin, Michele De Lucchi, and Georges J. Sowden), and of which he was the brilliant spokesman. In 1985, he went back to his agency Sottsass Associati, founded in 1981, and took up architecture again.

his style
Ettore Sottsass's productions have marked 1980s design. Denouncing industrial design as too impersonal, this admirer of the American Beat poets and the hippy movement creates with panache a warm, baroque, and insolent style. The bright colors and strange appearance of his deconstructed furniture, with its zigzags, stripes, and zebra patterns, are in sharp contrast to the sense of proportion advocated by functionalism. The balls and other amusing patterns that decorate them give the appearance of a construction game and pull us back into the toy boxes of our childhood.

best of
1967 Super Box furniture piece ◦ **1967** Menhir, Stuppas, Hydrants (ceramics) ◦ **1969** Valentine typewriter, Olivetti ◦ **1981** Claustra Carlton bookshelves ◦ **1981** Casablanca storage unit, Memphis ◦ **1981** Beverly buffet ◦ **1994** Esmeralda (Manufacture de Sèvres) ◦ **1994** Librairia bookshelves

Traforato basket, Alessi, 1982.

Adjustable seat, Olivetti, 1973.

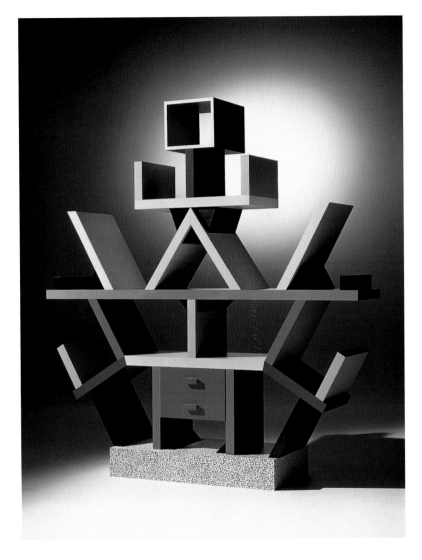

Claustra Carlton bookcase, Memphis, 1981.

He said:
"Existential curiosity, the desire to know other cultures, the desire, in short, to look over the garden wall, that's what pushes me to go further."

PHILIPPE STARCK

who is he?

Born in Paris in 1949, Philippe Starck studied at the École Camondo. Very quickly, his works attracted attention because of their cavalier treatment of the French art deco tradition. In the early 1980s, the design of Café Costes in Place des Innocents in Paris gave him a reputation which has not stopped growing since. The Royalton Hotel in New York and many other designs for public spaces were to follow. Yet this conceptual genius, now a star of international design, puts as much conviction into the design of a toothbrush for Fluocaril as a television for Thomson. His aim: to design products for everyone.

his style

The eclectic Starck style is happy to re-examine the classics that haunt our memory. But it breathes into them an astonishing modernity by reinterpreting them in contemporary materials: his Louis Ghost armchair and his La Marie chair, two magnificent examples of reuse, testify to an iconoclastic and extraordinarily inventive spirit. Each object that he ushers onto center stage plucks at our heartstrings: he accompanies their appearance with a breathtaking verbal demonstration and envelops them in a new legend. A rouser of conscience, Starck also fights to save the planet and, through his Good Goods operation, in assocation with the French mail order firm La Redoute, encourages the buyer to restrain his thirst for consumption. Between low profile and grandiloquence.

best of

1984 Designs the Café Costes, Paris ⬦ **1988** Designs the Royalton Hotel, New York ⬦ **1989** Fluocaril toothbrush ⬦ **1994** Televisions for Thomson ⬦ **1999** Good Goods mail-order catalog, La Redoute ⬦ **2000** First BON restaurant in Paris ⬦ **2002** Retrospective exhibition at the Centre Pompidou in Paris

Napoléon and Saint-Esprit, Kartell, 2000.

Kong restaurant, Paris, 2003.

Juicy Salif, Alessi, 1990.

He said:
"Every production produces love and tenderness, and I won't hide that this is what interests me most."

MARTIN SZEKELY

who is he?

Born in Paris in 1956, Martin Szekely studied at the Estienne school and the Boulle school in France. He began by designing manifesto furniture pieces to be shown at the Neotu Gallery. Then he entered the Galerie Kreo clique. He turned his creativity to craft techniques: pottery in Vallauris and glass at the CIRVA in Marseille, before making the move to industrial design, responding with precision to various commissions.

his style

The son of artists, from the outset Martin Szekely gave his works a personal touch very different from the contemporary codes. Between design and sculpture, his P1 chair, a masterpiece of balance and harmony produced in 1982, daringly associates straight lines, diagonals, curves, and semicircles. The work of a purist who knows how to play with the vocabulary of basic forms. However Szekeley succeeds in moving, gently and with great rigor, from the exceptional design of luxury goods to that of the most common objects. If his champagne bucket for Dom Pérignon or his dishes produced by Kréo are the reserve of a privileged few, his Perrier water glass and his comfortable Love Seat for the MK2 Bibliothèque cinema in Paris are accessible to everyone. As for his unexpected foldable chest of drawers, it mixes astuteness and technology with a touch of humor.

best of

1982 P1 chair ✣ **1987** Selected designer of the year at the Salon du Meuble, Paris ✣ **1996** Perrier glass, Cristal d'Arques (agent Kreo) ✣ **1999** L'Armoire, Galerie Kreo ✣ **2000** Des Plats, CIRVA, Galerie Kreo

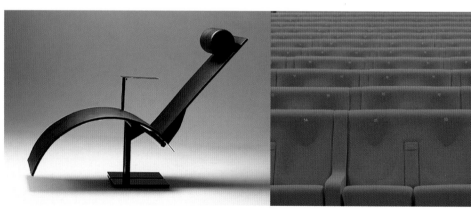

P1 chaise longue, Neotu, 1982.

Love Seat, movie theater seat for two, injected, covered foam, Kreo agent, MK2, 2003.

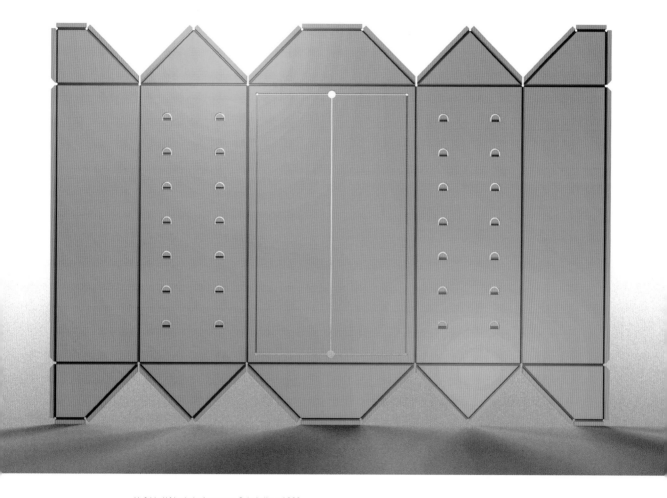

Unfolded Wardrobe in orange, Galerie Krea, 1999.

He said:
"My ambition is to find an economical effect that cannot be qualified as minimalist.
A commonplace."

ROGER TALLON

who is he?

An internationally famous designer, Roger Tallon trained as an engineer. His professional career developed hand-in-glove with consumer society and followed the exponential curve of the technological progress that marked the second half of the twentieth century. He worked for different design and planning offices and taught from 1957 at the École des Arts Appliqués in Paris. Unconcerned by notoriety, he nevertheless designed an impressive number of products. And his name remains forever synonymous with the famous Portavia television, a watch for Lip, and the TGV (high-speed French train). The Eurostar train and the Météor metro line in Paris also owe their fluid lines and spacious interiors to him.

his style

Tallon defines himself as an industrial designer. And nothing else. His passion: resolving problems. Whether it be for the Corail train, the RER train, the TGV duplex train, or Mexico City's metro, their interior design necessitated hours of discussion with dozens of professionals and millions of calculations: "Deep down," he says, "I am more of a mathematician than a designer." However he successfully touched upon the design of furniture and objects. Today, his Module 400 series of chairs and tables drive collectors crazy, and his folding TS chair and spiral staircase, designed for Sentou in the late 1970s, are still successful.

best of

1963 Portavia 111 television, Téléavia ❖ **1965** Module 400 series of chairs and tables ❖ **1967** 3T, crystal glasses, Daum ❖ **1967** Stainless steel cutlery for Ravinet d'Enfer ❖ **1969** Mexico City metro ❖ **1973** Mach 2000 watch, Lip ❖ **1981** TGV ❖ **1986** TGV Atlantique ❖ **2000** Météor

TGV Atlantique (high-speed French train), 1986.

Zombie armchair, 1967.

Bridge chair, module 400, 1965.

He said:

"Only mathematicians can appreciate design."

PATRICIA URQUIOLA

who is she?

This designer from the Basque Country in Spain, who has joined the front line of design stars, began studying architecture in Madrid. She finished her studies in Milan where she obtained her PhD thesis on the theme of the home-system with Achille Castglioni. She then became his assistant at the university before beginning a collaboration with Maddelana de Padova researching design and architecture which was to last five years. During this time, Patricia Urquiola designed the interiors of hotels, cafés, and shops with growing success. She then worked independently, designing furniture and objects, and also in collaboration with Piero Lissoni and Martino Berghinz.

her style

"I think that my formal register lies in sensuality. This seems normal to me. But I also fight for rigor in the way that I work." Patricia Urquiola's analysis of her own style aptly sums up the mixture of feminine and masculine in her works: rounded edges and widened forms lighten the powerful masses out of which her works seem to be carved. Intuition, warmth, and humanity envelop a solid structure which invites the body to make itself comfortable without losing itself completely. Her Fjord armchair covered in synthetic rubber, her Fat-Fat poufs, and the famous Lowseat are ideal versions of chairs.

best of

2002 Malmö couch, Moroso ✴ **2000** Lowseat couch, covered by fabric painted by Michael Lin, Moroso ✴ **2002** Fjord collection, Moroso ✴ **2002-04** Fat-Fat collection: pouf, bed, tables, B8B Italia ✴ **2005-06** Smoke couch and armchair, Moroso ✴ **2006** Antibodi, Moroso ✴ **2006** Prisma, Paola Lenti (with Eliana Girotto) ✴ **2006** Shangai Tip, Moroso ✴ **2006** Wallpaper Orize for the Tufty-Time couch, B8B Italia

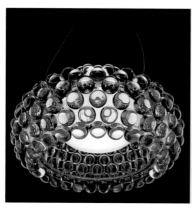

Caboche, Foscarini, 2005.

LOG2, Artelano, 2006.

Nido e Crochet, Paola Lenti, 2006.

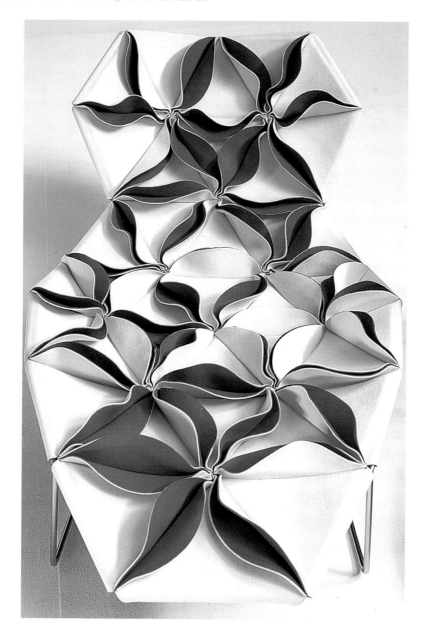

Antibodi, Moroso, 2006.

She said:

"Judge me by my objects, not by my ideas!"

MARCEL WANDERS

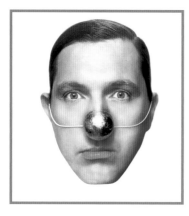

who is he?

This unpredictable leader of Dutch design was born in 1963 in Boxtel, and studied at the Academie Industriële Vormeving in Eindhoven and then at the Acedemie Voor Toegepast Künsten in Maastricht. A founding member of the Waac agency, he then founded his own agency, Wanders Wonders, in Amsterdam where he welcomed iconoclastic designers interested in experimental research. Wanders, however, is focused as much on the artistic gesture as on constructive collaboration with big manufacturers. He is therefore also director of the Moooi agency, which numbers KLM, Swatch, Apple, Rosenthal, Cappellini, Boffi, and Flos among its clients. These firms appreciate the talent and responsiveness of this ace of visual communications.

his style

Can we speak of a Wanders style? It is difficult to define, for the simple reason that when placed side by side the productions of this prolific designer show a lack of formal unity. Let's say they share a common spirit: the ability to provoke surprise. And admiration as well. Because, in the end, the works of this mixer of materials and sophisticated technologies tell stories as old as the world. His Zeppelin chandelier, enveloped in opaque plastic, seems to have come straight out of *The Fearless Vampire Killers*, whereas his rope Knotted chair, held together by its carbon-fiber frame, seems to have been put together by a fisherman. Playing with appearances is what this mischievous designer enjoys.

best of

1996 Knotted chair, Cappellini ❖ **1997** Spongevase vase ❖ **2001** Crochet table, Moooi ❖ **2005** Zeppelin light, Flos

Crochet Table, Moooi, 2001.

Stone Tourmaline, Kartell, 2006.

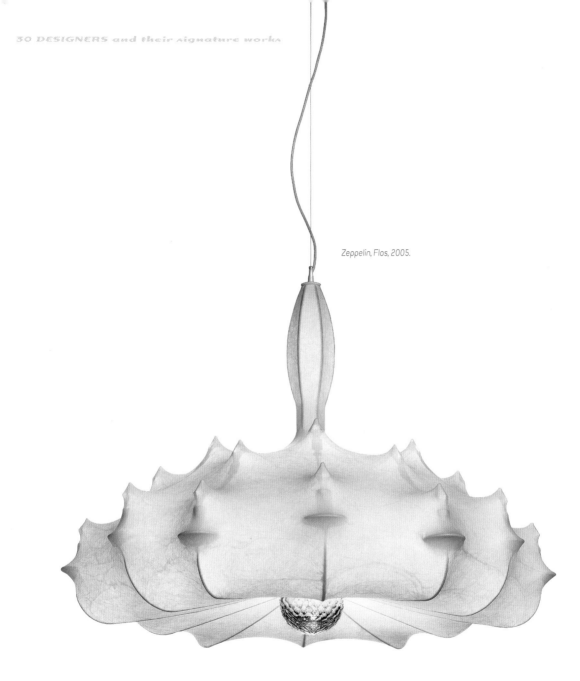

Zeppelin, Flos, 2005.

He said:
"Designers are there to change things, to find new structures and new ideas, but we would like to reach grandfathers as much as young women, and only the industrialist knows how to do this, because he ensures reliability, durability, and continuity."

Évolute lamp, Matali Crasset, Danese, 2004.

APPENDIXES

A SHORT GUIDE TO FRENCH STYLE

Design is, and always has been, an internationally evolving concept, with designers borrowing from one region or decade to the next, resulting in immense variety. We have chosen French design to illustrate the historical progression of one of the world's most familiar and celebrated styles. Over the course of history, kings and queens have marked their reign by commissioning great works or encouraging the creation of a new style of furniture. Straight or curved lines, richly ornamented or not, references to antiquity or nature, exotic or local wood, geometric or floral designs, excess or simplicity, all constitute the grammar of each stylistic period: here is a survey.

Louis Ghost armchair,
transparent polycarbonate,
Philippe Starck, Kartell, 2002.

DARK AND RIGOROUS: Louis XIII

A little history

This moment in history was marked by the settling of society, reflected in the appearance of fixed furniture filling the role of architectural elements. It replaced chests, trestle tables, and folding chairs in mansions and stately homes. However, although the reign of Louis XIII extended from 1610 to 1643, the style bearing his name covers a longer period. It began under Henri IV and Marie de' Medici around 1600, and continued until the death of Mazarin in 1661.

What does it look like?

The general appearance of Louis XIII furniture is somewhat massive. The transformation took place gradually. The influence of the Italian Renaissance style, which found full expression in the building of the châteaus of the Loire valley and especially the Fontainebleau château in 1528, diminished and the mannerism of the Henri II style slowly gave way to a new classical rigor. Cohabitation reigned: ebony cabinets or tables inlaid with colorful bouquets, in keeping with the Dutch fashion, went well with the somber, diamond-point sculpted, double sideboard. With the armchair, they symbolize today the solid and somber Louis XIII style which is also marked by skillful lathe work: thanks to the perfecting of lathing techniques, columns, legs, arms, and braces are richly decorated with coils and twists.

1

How to recognize it

❋ Seat legs and armrests, and table or dresser legs sculpted in twists, strings of beads, and balusters
❋ Legs that end in a round or square shape
❋ Molded feet or balls decorating the intersection of chairs or tables
❋ H-shaped struts
❋ Diamond-point relief
❋ Inlaid colored bouquets
❋ Chairs stuffed with horsehair and covered in leather or tapestry
❋ Slightly incurved armchair armrests ending in crosses

Key pieces

❋ Double sideboard
❋ Armchair
❋ High-backed armchair
❋ Ebony-veneered cabinet

3

2

1 Double sideboard with doors, seventeenth century.
2 Double sideboard, seventeenth century.
3 Double-headed chair, seventeenth century.

RICH AND POMPOUS: Louis XIV

A little history

The Louis XIV style symbolizes the reign of the the Sun King. The style was born in Versailles, which began to be enlarged in 1669, and it reflects the château's sumptuous majesty. It is laden with heavy decorations in the image of this absolute monarch, whose influence in Europe lasted over a quarter of a century. In charge of this incredible building site was the painter Charles Le Brun, who invented the great Louis XIV style. However, from 1680, a parlous state of finances brought the extravagant spending to an end, and the Louis XIV style entered a less luxurious and less pompous phase.

What does it look like?

The Louis XIV style is recognizable by its sense of pomp: silver furniture decorated the Hall of Mirrors for twenty years before being melted down and the proceeds used to swell the coffers of the treasury. Originally from Italy, the commode or chest of drawers appeared in the king's bedroom in the form of a pair of commodes inlaid with copper and tortoiseshell and enhanced with gilded bronze. Consoles, tables, cabinets, armchairs, and candelabras are sculpted in gilded wood. They feature ostentatious ornaments including shells, acanthus leaves, trophies, and squared patterns, as well as gilded and engraved bronze fittings.

1

How to recognize it
* H-shaped struts developing little by little toward an X-shape
* Intricately sculpted furniture legs
* Incurved armrests ending in a cross or a scroll
* Table edges inlaid with fantastical, sculpted decorations
* Gilded bronze hinges, door handles, entrance locks, and corners
* Curved leg tops

Key pieces
* Sculpted and gilded console covered in marble
* Chest of drawers inlaid with copper and tortoiseshell
* Eight-legged desk known as the Mazarin
* Pedestal table covered in fabric

1 Desk belonging to Maximilian-Emmanuel, elector of Bavaria, attributed to André-Charles Boulle (1642–1732), c. 1715.
2 Eight-legged Mazarin desk, double-strutted and cased in Boulle's style. Inlaid with copper and inlaid with red tortoiseshell.
3 Mazarine chest of drawers inlaid with brown tortoiseshell and copper against an ebony background, gilded and engraved bronze, cherry marble, made in c. 1708–09 for the bedroom of Louis XIV at the Grand Trianon Palace, André-Charles Boulle (1642–1732).

2

3

COMFORTABLE AND REFINED: Louis XV

A little history

Seen as the pinnacle of refinement and skill, the Louis XV style covers a period spreading from 1715 to 1774. It began with the reign of Philippe d'Orléans and ended with the king's death. An unprecedented intellectual effervescence characterized life in Paris where salons held by aristocrats were organized around competitions in wit and curiosity. The period's top philosophers, writers, poets, and intellectuals gathered there. During this half-century, three stylistic movements succeeded each other, marking the transition from the curved to the straight line: the Regency style, the rococo style, and the Transition style. During the Regency style, the court moved to the Palais Royal in Paris. Then, under Louis XV, Versailles rediscovered its splendor of old. However, thanks to Mme de Pompadour, who was crazy about decoration, the charm of private apartments was discovered. Toward 1750, archaeological excavations in Italy uncovered the ruins of Herculaneum, which triggered a return to antiquity and to simpler, less elaborate forms.

What does it look like?

Generous volumes and harmonious lines mark the Regency style. It developed during the refurbishment of town houses situated around the Place Vendôme in Paris, the haunt of courtiers. For the first time, comfort is a priority. The famous cane-seated or tapestry wing chair follows women's fashion closely: its seat was widened and its armrests moved apart so that the queen and her attendants could sit down in their voluminous hoop skirts. More functional than the previous style, Louis XV furniture still had an important decorative role. It can be recognized by its sinuous lines and the sometimes exuberant sculptures which decorate the furniture, reflecting the period's lightheartedness and fantasy. The return to more sober and masculine forms will later significantly modify furniture's silhouette.

1

2

1 Louis XV cylinder writing desk in Versailles, Jean-François Oeben (1721–1763) and Jean-Henri Riesener (1734–1806), 1760–69.

2 Salon table with inlaid wood and gilded bronze, seventeenth century.

3 Cabriole wing chair (one of a pair), stamped Louis Delanois (1731–1792), c. 1765.

How to recognize it

- Veneering that plays on the visual effects of the wood grain
- Gilded bronze
- Furniture with pull-out shelves, secret drawers, pull-out flaps, etc.
- Ornaments inspired from the plant word: flowers, fruits, leaves, and twigs
- Painted Chinese patterns
- Curved legs

Key pieces

- Rounded chest of drawers with curved legs
- Desk with a leather top and three belt drawers
- Paneled wainscoting
- Caned armchair
- Sloping or folding writing desk
- Numerous little tables for various activities

3

NEOCLASSICAL AND ELEGANT: Louis XVI

A little history

A period of contradictory effervescence. On the one hand, the imagination of intellectuals was excited by technical and scientific progress at a time when the American civil war and far-off expeditions were bringing to light new civilizations. On the other hand, fashion was extolling a classical aesthetic which had regained public popularity thanks to the numerous archeological excavations in progress. In face of a changing world, aristocratic society was searching for a reassuring sense of continuity in these ancient cultural roots.

What does it look like?

The birth of the Louis XVI style, called neoclassical, preceded the king's coronation in 1774 by a few years. Its beginnings correspond to the break caused by the "Greek" fashion launched around 1755 by the Marquise de Pompadour's brother, the Marquis de Marigny, director of the king's buildings and manufacture. Marie-Antoinette softened this new, somewhat rigid style with the decoration of the Petit Trianon Palace in the grounds of the Château de Versailles and its numerous small pieces of furniture: armchairs with round or oval backs, canopy beds with cut panels, and so on. Likewise, the decoration of the Folie Bagatelle by the Count d'Artois, on the edge of the Bois de Boulogne on the outskirts of Paris, opted for more elegance and lightness while keeping hold of the fundamental rigor of the Louis XVI style.

How to recognize it

❋ Straight and round lines
❋ Grooved, spindle, or tapered legs
❋ Horizontal armrests
❋ Light lacquered wood
❋ Simplified ornaments
❋ Wallpaper fashion
❋ Blue or white marble
❋ Lacquered furniture with Japanese patterns

Key pieces

❋ Light cabriole chairs with "en medallion" or "à chapeau" backs
❋ Backrests hinged to the seat
❋ Round or oval dining-room tables, mounted on wheels
❋ Small round or oval tables edged in copper

❋ Wedding chests with doves on the front and bayleaf friezes

1 Chest of drawers decorated with porcelain plaques (Manufacture de Sèvres), belonged to Mme du Barry, stamped Martin Carlin (c. 1730–1785), c. 1775.

2 Round mahogany dining table, late eighteenth century.

3 Chair, apartment of the Dauphin, Château de Compiègne, Georges Jacob (1739–1814), 1791.

SIMPLE OR OSTENTATIOUS: the Empire style

A little history

Between the French Revolution in 1789 and the coronation of Napoleon and Josephine at Notre-Dame Cathedral in Paris in 1804, France underwent major political change. During these chaotic years, the country lived through the passage from the old to the new regime with exaltation but also in blood and tears. The execution of Louis XVI and Marie-Antoinette, and the rise to supreme power of an army general just ten years later marked the emergence of a new society which broke with the past. Paradoxically, in the decorative arts, save for a few fads and incidental adjustements, a certain conservatism reigned at the heart of the triumphant haute bourgeoisie.

1

What does it look like?

The beginnings of the Empire style appeared during the Directory period, just after the Egyptian campaign of 1798. The decoration of the Malmaison, entrusted in 1800 by Josephine de Beauharnais to Percier and Fontaine, shows the influence of Egyptomania on furniture and decoration. But it also continued the spirit of the previous style: classicism, the use of mahogany, and simple forms (adapted to the recent mechanization of manufacturing). However, at the high point of Napoleon's reign, this style took on a certain sense of pomp which paid tribute to imperial ostentation and its symbols through a rich ornamentation of eagles and bees. A pronounced taste for the classical, the Roman, and the Pompeian developed under the aegis of the painter David.

Key pieces

* Mahogany beds with high, flared edges
* Chairs with perforated backs
* Mahogany consoles
* Antique-style curule chairs
* Bedside tables

1 Early nineteenth-century pedestal table which can be used as a dining table or a living-room table.

2 Pair of wingchairs with curved front and armrests with the head of an Egyptian woman, and clawed front feet, delivered in 1805 for Madame Mère's furniture in the Salon des Glaces in the Grand Trianon Palace, François-Honoré-Georges Jacob-Desmalter (1770–1841).

3 Chest of drawers, early nineteenth century.

How to recognize it

* Armrests stop short (flowing dresses are in fashion)
* Discreet sculptures and friezes
* Classical ornamentation with sibyls, sphinxes, swans, or palm branches
* The armrests continue down to the chair legs

2

3

BOURGEOIS OR THEATRICAL: Louis-Philippe

A little history

After Napoleon's fall in 1814, and up until the arrival of Louis-Philippe on the throne in 1830, France experienced turbulent times. Louis XVIII, then Charles X marked the return of the Bourbons to government. But this restored, revanchist monarchy was addicted to the past, and collided with the spirit of progress afoot in Europe. As in England, industrialization gave rise to a new social landscape. The bourgeoisie became the dominant model. And when he ascended to the throne, Louis-Philippe adopted a comfortable lifestyle, dipping into the reserves of Restoration furniture which had already begun to soften the Empire lines. Meanwhile, under Napoleon III, the Empress Eugénie was influential in the return to fantasy in matters of decoration. Baroque and shimmering, it borrowed from various styles.

What does it look like?

The Louis-Philippe style corresponds to the need for intimacy and comfort of well-to-do, urban customers who lived in the new quarters of the big cities. It was perfectly adapted to cabinetmaking's new mode of mechanical production. It also reflected a nostalgia for past forms. Toward 1827, the fashion of Renaissance furniture made the fantasy of the Henri II sideboard an essential part of affluent dining rooms. The Napoleon III style with its stuffed low fireside chairs, heavy double curtains, delicate little black chairs, plump poufs, and exotic green plants, brings a certain feeling of theatricality into everyday decoration which goes back to the lightness and elegance of eighteenth-century salons.

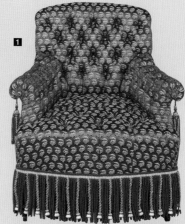

How to recognize it
- Dark wood with light inlays
- Curved backs
- Crossed, scrolled, or swan-necked armrests
- Cuffed armrests
- Square, curved, or frog-legged legs
- Padding, tassels, fringes, and lambrequins

Key pieces
- Mirrored wardrobes
- Tub armchairs
- Hardened pasteboard chairs
- The gondola chair
- Companion chair
- Henri II chairs and dressers loaded with sculptures
- Voltaire chair

1 Second Empire chair, Napoleon III (1852–70), Fontainebleau.

2 Neo-Gothic pedestal, after Louis Charles Théodore Charpentier (1797–1867), attributed to Michel-Victor Cruchet (1815–1877), c. 1835–37.

3 Armchair from a set of eight, Michel-Victor Cruchet (1815–1877), Louis-Philippe (1830–48), 1847.

4 Armchair, one of a pair, Claude-Aimé Chenavard (1789–1838), c. 1835.

INSPIRED BY NATURE: art nouveau

A little history

The transition from the nineteenth to the twentieth century contains the seeds of the modern period. During the Belle Époque, industrialization picked up pace and International Exhibitions, starting in 1851, were a privileged means of showing both technical innovation and the discoveries resulting from colonization. The invention of electricity, and then of cinema in 1895, as with the first cubist paintings in 1908, violently upset people's habitual way of seeing things. Is this why nostalgic furniture became so successful? Reinterpretations or reinventions of the eighteenth century fascinated the bourgeoisie from the wealthy quarters, while copies of all kinds supplied most of the market. It was in this somewhat fearful context that art nouveau triumphed at the International Exposition of 1900.

What does it look like?

In 1882, the creation of the Union Centrale des Arts Décoratifs, and then of the school in Nancy, France, under the guiding influence notably of Émile Gallé and Louis Majorelle, contributed to the creation of a style that, in the face of increasingly mechanized manufacturing, returned to artistic cabinetmaking and glorified inlaid or solid wood. It is characterized by taut lines in floral scrolls, waves, and arabesques vaguely reminiscent of the rococo style. The "modern style" or *style nouille* combines virtuosity and elegance.

1

How to recognize it

⊛ Precious wood inlay
⊛ Gilded bronze for door handles or entrance locks
⊛ Armrests and legs forming a continuous line
⊛ Floral repertory with a preference for climbing plants: bindweed, wisteria, convolvulus, and sweet pea

Key pieces

⊛ The bed and bedside table
⊛ Nest of tables
⊛ Water-lily table
⊛ The pedestal table
⊛ Vases and lamps in glass paste

2

1 Lamps with umbel motifs, Émile Gallé (1846–1904), c. 1902.
2 Sloe pattern vase, Daum Frères, 1909.
3 Water-lily desk, Louis Majorelle, c. 1900.
4 Porte Dauphine metro station, Paris, dragonfly glass porch and enameled panels of lava stone, Hector Guimard (1867–1942), 1900–02.

4

3

KEY DATES IN THE HISTORY OF DESIGN

The history of design begins in the late nineteenth century with the first mass-produced furniture. Marked by diverse and extremely rich events, it is traversed by passionate and ideological fights which regularly divide its protagonists. Likewise, the domination of one avant-garde school, the formation of break-away movements, or the influence of an important exhibition represent major turning points at different periods. And create seminal moments in the development of taste, style, and behavior. Here is a list of key dates, each with a description, which help to define the saga of design.

Twelve 365,
Naoto Fukasawa,
Issey Miyake, 2005.

1851

Great Exhibition, Crystal Palace, London, England

This exhibition was the first industrial and technical show in history to bring together products and goods from around the world made according to all known methods. From then on, International Exhibitions would set the tone in style and fashion.

1862

Art & Crafts Societies, England

Founded by the architect William Morris, this movement, which included Charles Robert Ashbee, William Arthur Smith Benson, and Christopher Dresser among its members, aimed to bring together artists and craftsmen in order to make the general public's lifestyle more harmonious. Although originally vociferously opposed to the industrial manufacturing of furniture, certain protagonists of the movement, responsive to progress, would finish by accepting mechanization. However, they continued to insist upon the importance of designing specifically adapted forms.

1889

Darmstadt Artists' Colony
Duchy of Hesse, Germany

The Grand Duke of the Duchy of Hesse, Ernst Ludwig von Hessen und bei Rhin, a lover of the arts and a great patron, founded the Artists' Colony of in 1889 under the guiding influence of the architect Josef Maria Olbrich. He called upon seven artists (painters, graphic designers, architects, and interior designers) who believed in uniting art and production. The members of the colony defended the concept of the total artwork linking the artist and the craftsman, life and creation.

1894

Glasgow School, Scotland

When he was named director of the Glasgow School of Art, a school considered the most progressive in Europe at that time, Francis H. Newbury encouraged the development of a movement based on geometrical forms. He wanted to break with historicism and the excesses of French and Belgian art nouveau. The architect Charles Rennie Mackintosh formed a group, The Four, and promoted an original style whose straight and austere forms are still a reference.

1887

Vienna Secession, Austria

Director of the Vienna Academy's school of architecture, the architect Otto Wagner published a theoretical manifesto in 1895 entitled *Something That is Functional Can Never be Beautiful*, which later served as a basis for the radical Secession movement. In the name of free creativity and modernity, he defended the idea of total equality between the applied arts and the fine arts, emphasizing the virtues of simple, sober, and functional forms. A vision of art and architecture which broke completely with history and tradition.

1907

AEG employs Peter Behrens, Germany

The architect and painter Peter Behrens played a decisive role in the advent of modernity. His curriculum vitae is impressive: he was co-founder of the Secession in Vienna, part of the Darmstadt Artists' Colony, and directed the school of decorative arts in Düsseldorf. Reason enough for the AEG company (Allgemeine Elektricitäts Gesellschaft) to entrust him in 1907 with the company's whole image: the building of the factory, the worker's lodgings, the offices, shops, packaging, logo, graphic design, and advertisements as well as the arc lamps and electric teapots. An unprecedented example: the first milestone in global design.

1919

Bauhaus school, Germany

On April 1, 1919, Walter Gropius designed a concept for the merging of the Weimar Academy of Fine Arts and its School of Applied Arts into one institution: and so the Bauhaus school was born. Its aim was to create "a collaboration between the artist, the industrialist, and the technician, which if organized according to the spirit of the times, would be able to replace the factors of what previously had been individual work." A program that ushered in a revolutionary mindset. It was to result in passionate theoretical debates on the place of man in a mechanical environment and the creation of original forms adapted to the life and spirit of modern times.

1925

Pavillon de l'Esprit Nouveau (The Pavilion of the New Spirit), Paris, France (International Exhibition of Modern Industrial Applied Arts).

The International Exhibition of Modern Industrial Applied Arts which took place in Paris in 1925 shows many prestigious pavilions created by great designers (J. E. Ruhlmann, L. Sue, A. Mare). The Pavillon de l'Esprit Nouveau, designed by Le Corbusier, Ozenfant, Amédée, and Pierre Jeanneret, was situated on the fringes of the exhibition, however, and stood out from the crowd. Conceived and designed according to an anti-decorative program, it emphasized the advantages of prefabrication and austere furniture with forms perfectly adapted to their function.

1926

The First Salon des Arts Ménagers, (Domestic Science Fair), Paris, France

The Salon des Arts Ménagers in Paris exhibited the first consumer goods that were to change the housewife's life: washing machines, ovens, flat irons, vacuum cleaners, etc. It took place shortly after the war at the time of the first stirrings of consumer society and was an immediate success. An attraction in itself, it was visited by millions of people.

1929

UAM (Union des Artistes Modernes), Paris, France

The Union of Modern Artists was born on May 15, 1929. A small group of innovators, René Herbst, Djo-Bourgeois, Jean Luce, Jean Puiforcat, Georges Fouquet, and Charlotte Perriand, decided to split off from the conservative fringe which they reproached for its inertia. They participated in the 1929 Salon d'Automne with the intention of "bringing together artists of the same movement and state of mind, of grouping their efforts, and insuring visibility by organizing an annual international exhibition in Paris and a propaganda bulletin." In 1934, with the aim of forging closer links with industry, the UAM, presided over by Robert Mallet-Stevens, created a permanent commission of exchange with the OUA (Office de l'Utilisation de l'Acier—Office of Steel Usage) in order to work with building materials such as metal, glass, and rubber. Works by Pierre Chareau, Jean Prouvé, Louis Sognot, Francis Jourdain, and René Coulon, amongst others, testify to this taste for experiment and invention.

1930

Streamline Movement, United States

Popular in the United States during the 1930s, the Streamline style boosted mass consumption after the 1929 depression. American designers developed a smooth and streamlined aesthetic, inspired by research in aerodynamics, speed, and movement. The Streamline sowed the seeds for an optimistic vision of society in the future. The famous K4 locomotive redesigned by Raymond Loewy in 1936, for the Pennsylvania Railroad, is one of its most famous emblems.

1932

Modern Architecture: International Exhibition, MoMA, New York, United States

Organized at the Museum of Modern Art in New York by Philip Johnson and H. R. Hitchcock, this exhibition presented productions with clean lines which shared a common aesthetic qualified as "The International Style," represented by the architects Mies van der Rohe, J. J. P. Oud, Le Corbusier, and Frank Lloyd Wright. The same year the museum was the first ever to inaugurate a department of architecture and industrial Art.

1937

New Bauhaus, United States

In face of the rise of Nazism, which put the school's very existence into question, the Bauhaus was obliged to change its location twice. First to Dessau in 1925, and then to Berlin in 1931, until it was closed down permanently by the police on April 11, 1933. But the Bauhaus school rose from its ashes on the other side of the Atlantic. Founded in 1937, the New Bauhaus opened its doors in Chicago with the support of the local Association of Arts and Industries. The spirit was the not the same however. Against the patrons' wishes, its director Moholy-Nagy was determined to bring back experimental design and in 1933 founded the School of Design, which became the Institute of Design in 1944. These schools were to have a considerable influence on the new generation of American designers and architects.

1940–41

Exhibitions of the Industrial Design Department of MoMA, New York, United States

Several exhibitions organized by the Industrial Design Department of the Museum of Modern Art in New York, in particular *Organic Design Home Furnishing* and *Low Cost Furniture Design Competition,* transformed the United States into a hub of international design. They gave designers an important role in the emergence of the American way of life. The competitions set up on these occasions favored experiments in techniques and materials resulting from the war industry.

1942
Utility Products exhibition, England

In 1942, war was raging across Europe. In England, industry was at the service of war, and the civilians were subject to restrictions and rationing. Hugh Dalton, a high-ranking civil servant at the Ministry of Trade, had the idea of creating a committee for "utility products." The manufacturing of a limited number of indispensable products was authorized, those used in everyday life: clothes, crockery, furniture, etc. Built out of good quality materials, they were made to last, although their basic colors and simple shapes meant that they still corresponded to the criteria of modernity.

1943
Founding of the Ikea stores

The Swede Ingvar Kamprad founded Ikea in 1943. Originally specialized in the mail-order selling of stationery, he began selling furniture in 1947, and very soon had thought up a way to manufacture inexpensive, mass-produced furniture that could be distributed according to a innovative new commercial system. He launched a catalog in 1951, forged links with designers, and in 1956, began to specialize in furniture in kit form. Situated on the outskirts of cities, Ikea stores offered a large range of colorful utensils, objects, and furniture in simple shapes. A combination which still works today: the brand can be found in twenty-eight countries at present.

1947
The Ulm school, Germany (Collaboration with the Braun firm)

This private school of applied arts (Das Hochschule für Gestaltung d'Ulm), inaugurated in 1955 and directed initially by the ex-Bauhaus student Max Bill, has marked the history of design. Familiar with scientific and technological advances and in phase with new modes of manufacturing and distribution, over the years the teaching gradually moved toward industrial design. The collaboration between Hans Gugelot,

in charge of the school's design-product department, and the appliance brand Braun, is an example of the sober, functional aesthetic which brought Braun so much success.

1949
Useful forms, objects of our time
Paris, France

The "Formes Utiles" (Useful Forms) movement created in 1949 originated in the UAM. The idea was to place industrial and craft productions under the same name, not out of nostalgia for the past, but out of a desire for balance between progress and man. André Hermant wrote in the 1955 manifesto: "We name as useful the form of an everyday object, of a building when it is the exact expression of a correspondence between efficient use, economical materials, and the satisfaction of sensitivity and intellect."

1951
The triumph of Scandinavian design
at the Milan Triennial

Founded in 1923, the Milan Triennial is an important reference. It contributed to the international recognition of Scandinavian design, which it promoted for many years. In 1954, the exhibition "Design in Scandinavia," which was later shown in several cities across the United States, increased the commercial success of Scandinavian designers such as Hans Wegner, Arne Jacobsen, Poul Kjaerlholm, Pirkko Stenros, and Saara Hopea.

1964
Founding of Habitat, England

Habitat was an innovative concept in stores, founded in London in 1964 by the designer Terence Conran in response to the demands of a new generation and their desire to adopt a fresh way of living. Conran offered designer products (furniture, objects, utensils, and accessories) adapted to a non-conformist state of mind, inviting customers to customize their decors.

1966
Archizoom and Superstudio, Milan, Italy

As early as 1966 in Florence at a time of great debate, the architect, designer, and theoretician Andrea Branzi founded—with Gilberto Corretti, Paolo Deganello, and Massimo

Morozzi—the first group of Italian avant-garde critics: Archizoom Associati. Proclaiming an iconoclastic practice, associating architecture and other arts, Archizoom repudiated the generally accepted concept of design at the service of the economy.

1967

Seventh Furniture Fair in Milan, Italy

The Milan Fair of 1967 was dominated by bright colors and pop art patterns. Plastic furniture triumphed. This light, practical, washable, unbreakable, bright, and colorful material was a symbol of the carefree attitude of the 1960s generation. It inspired flexible and ergonomic shapes. Enthusiastic and ready to take risks, many Italian furniture manufacturers were already transforming Italy into a hotbed of design.

1969

Founding of the Centre de Création Industrielle (CCI), Paris, France

Under the auspices of Georges Pompidou and André Malraux, design made an institutional breakthrough in France with the founding in 1964 of the Atelier de Creation Contemporaine (Studio of Contemporary Creation). An initiative which was followed in 1969 by the birth of the Centre de Création Industrielle. The role of this new department of the Union Centrale des Arts Décoratifs was to select products sold on the French market, to document their technical information, and organize exhibitions on the different aspects of design.

1969

Exhibition of Modern Design in Japan, Museum of Modern Art, Kyoto, Japan

Japanese design, influenced for a long time by American design, began to specialize with the commercial success of firms such as Sony, Mitsubishi, Hitachi, and Matsushita—firms which employed designers to give their transistor radios, video recorders, cameras, and stopwatches simple, sober, and modern shapes. But the exhibition in Kyoto shows that Japanese design is more than simply industrial design. Isamu Noguchi's lamp and works by Saburo Inui, Kenzo Tange, and Daisaku Cho revealed a surprising capacity for combining tradition with modernity.

1972

Italy, the New Domestic Landsape, exhibition at MoMA, New York, United States

Held at MoMA , the exhibition acknowledged the vitality of Italian design. It went beyond the simple display of products and asked different designers to create environments showing design's recent developments. Young Italian designers were therefore invited to come up with a domestic environment which took into account new ways of being. The result was two conflicting tendencies: one which privileged a design based on comfort, and the other, more radical, which upset our everyday habits and criticized design itself.

1974

Global Tools, Milan, Italy

Born in Florence in November 1974, as part of the anti-culture and anti-design movement, Global Tools was an anti-school of architecture and design, bringing together protagonists of the radical Italian avant-garde around the same revolutionary ideas. New groups such as UFO, 9999, Studio 65, and Gruppo Strum emerged. The experiment lasted a year but is still influential, and is at the origin of works influenced by pop art and arte povera.

1976

ICSID *Design for Need*, London, England

The oil crisis put things straight. Productions had be made with less expensive materials. The ICSID organized a conference in London on the theme of economizing energy: *Design for Need*. Designers were sensitized to the problems of over-production and the positive role they had to play in the fight to save the planet. Recycling methods were put forward as a priority.

1979

Creation of the VIA, Paris, France

The VIA (Valorisation de l'Innovation dans l'Ameublement–Valorization of Innovation in Furnishing) had the mission of promoting and defending French design in the domain of furnishing (furniture, lights, tableware, decorative accessories, rugs, furnishing fabrics) both in France and other countries. Supported by the furniture industry, it was innovative in its financing of prototypes and its connecting of designers with potential manufacturers.

1979

Studio Alchimia, Furniture Fair, Milan

Created in Milan in 1976 by Alessandro and Adriana Guerriero, who were then joined by Alessandro Mendini and Ettore Sottsass, the Studio Alchimia group was a hit at the Furniture Fair in Milan in 1979. The exuberant aesthetics of the multicolored and baroque furniture it presented broke with the hegemony of rational functionalism.

1981

Memphis, Italy

Founded in Milan by Ettore Sottsass and Michele De Lucchi, The Memphis group, whose theories were as provocative but less nihilistic than those of Studio Alchimia, welcomed designers of all nationalities and became a great melting pot of totally crazy creations that were shown in galleries, published in avant-garde reviews, and very quickly became collector's items.

1983

Creation of the Domus Academy, Milan, Italy

Andrea Branzi created the Domus Academy in 1983 in Milan. This school, a focal point of thought and discussion, attracted the best designers and enthusiastic students from across the world. The teachings of Richard Sapper, Ettore Sottsass, and Achille Castiglioni preached in favor of total creative freedom.

1989

Opening of the Design Museum, London, England

Now near the Tate Modern, the Design Museum has rapidly become a centre of consecration for the designers that are exhibited there. Designers that show at the Design Museum can be sure of an international reputation.

1993

Design, Miroir du Siécle (Industrial Design: Reflection of a Century) at the Grand Palais, Paris, France

In a desire to reduce waste, Philippe Starck presented two hundred models of basic indispensable objects in the La Redoute catalog. An act strangely reminiscent of the "utility products" operation undertaken in Britain during the Second World War.

September 2006

Reopening of the Musée des Arts Décoratifs in Paris, France

Closed to the public for several years, the Musée des Arts Décoratifs in Paris reopened its doors with a new interior design, making access to its splendid collection of 150,000 objects easier. Designer objects and products testify to the French *art de vivre* and the skill of professional craftsmen and manufacturers, and also to the research and creativity of living designers.

2007

The opening of the Cité du Design in Saint-Étienne, France

The Design Biennale, which has taken place in Saint-Étienne since 1993, is today an absolute must for design professionals, attracting an ever-growing public. The opening of the Cité du Design is set to turn the old manufacturing town of Saint-Étienne into an international capital of design, and confirm the strength of French design today.

GLOSSARY OF KEY DESIGN TERMS

Design has its leaders and its lovers. They share a common vocabulary, which distinguishes them from the rest of the crowd filling the fairs, biennales, and other exhibitions devoted to this discipline apparently accessible to everybody. Doesn't it concern objects devised simply to embellish our immediate surroundings? But design already has a long history. And its field of application grows according to technical innovations and new ways of behaving. Its grammar and vocabulary are continually enriched and constitute a body of references which separates the insiders from the others. It is impossible to be "one of them" if you do not know the name of the famous school of applied arts that marked the advent of modernism: the Bauhaus. It is also important to be able to distinguish a logotype from an icotype. To help you keep up to date and not be left behind, here are a few important notions of which you should be aware.

*Noli Me Tangere,
polyurethane
foam poufs,
Vincent Beaurin,
VIA edition, 1994.*

B

BAUHAUS

School of art and architecture founded in Weimar, in Germany, by Walter Gropius in 1919. It contributed to the creation of a totally new aesthetic bringing together art and technique. Several of the greatest designers of the early twentieth century taught there, including Paul Klee, Josef Albers, Wassily Kandinsky, Laszlo Moholy-Nagy, Ludwig Mies van der Rohe, Oskar Schlemmer, and Johannes Itten.

BRAND

More than just a trade name, a brand also serves to identify a firm or a service. It gives value to a product and constitutes a firm's capital. A brand can be expensive.

BRAND IMAGE

The brand image of a product is built up over the years. It corresponds to notions such as strength, honesty, and reliability, and creates a relationship of trust between the consumer and the brand. When this is broken, by a defective series for example, it can take a long time to build up again.

BRANDING

Branding is the way in which a brand creates a scenario around its name, which conveys its values and enhances the product. For example, Nike's "Just do it" seems to say to each customer: "Go on, push yourself, you can do it!"

C

CAD

Computer-Aided Design

COMPOSITE MATERIALS

This term applies to a group of composite materials made from a plastic base reinforced with fibers. The plastic serves as a link between the fibers and guarantees the material's rigidity and resistance. The most commonly used fibers are fiberglass, carbon fiber, and Kevlar fiber. The mix is made according to two technical processes: thermo-hardening (polymers which harden under heat) and thermo-plastic (polymers which harden when cooled). The first is irreversible whereas the second is reversible, and so ideal for recycling.

CORIAN®

This is the favorite new material of designers. Made by Dupont de Nemours and regularly used for kitchens or bathrooms, its translucency, flexibility, and contemporary appearance makes it a hit with furniture designers. Obtained from a mixture of resistant, natural materials and acrylics, it can be sculpted and worked like wood.

Soleil coffee table by Jean-Michel Frank, Écart, 1930.

D

DESKTOP PUBLISHING

The production of printed matter by means of a desktop computer having a layout program that integrates text and graphics.

E

ERGONOMICS

An ergonomic form follows the shape of the body. Moreover, ergonomics is the study of work conditions, in particular the relationship between man and machine. It is a subject which calls upon sociology, social psychology, and anthropometry, among other disciplines. It physically introduces the consumer into the product's design process.

F

FAIR

Design has its fairs just like the fine arts. The most famous is the Furniture Fair in Milan which owes its reputation to the dynamism of Italian design during the 1960s, 1970s, and 1980s.

FUNCTIONAL

When the shape of furniture or an object is the result of its function, it is called functional. In the 1920s, the straight lines and austere rigor of functionalism represented a strong ideological commitment for the modernists. It accompanied their mission: to create a beautiful and inexpensive environment for everyone.

G

GLOBAL DESIGN

When a company entrusts an architect designer with the design of its buildings, the graphic design of its logo, brochures, letterhead, and packaging, we talk of global design. Peter Behrens was the first to be entrusted with such an important task by AEG (Allgemeine Electricitäts Gesellschaft) in 1907.

GRAPHIC CHARTER

A graphic charter, or guidelines, is a collection of graphic characters for the visual communication of a brand, company, or community. Used for all possible forms of communication.

I

ICOTYPE

A visual symbol which identifies a brand. Example: the Macintosh Apple or the swoosh of Nike.

L

LOGO

Abbreviation of the word logotype. It signifies the graphic representation of a commercial brand or company acronym. It is presented in the form of an abstract or figurative symbol. It must be immediately recognizable and resist time. Ferrari's horse or Vuitton's LV have conquered the planet. The puma on certain athletic shoes is internationally recognized, while the Coca-Cola bottle has became a classic.

LOW DESIGN

This term is used to define discreet, unpretentious design forms which reject a proliferation of signs. Low design goes against manifesto furniture which marked the transition from sober modernism to eclectic postmodernism in the late 1970s. Elegance and gentle lines characterize the furniture and objects with this label.

M

MARKETING

A widely used term. The way in which a company analyzes the market, and positions itself in relation to it, investing the different levels which separate it from the customer. Marketing is concerned with commercial advertising, salesroom promotions, the training of salesclerks, and the search for new products.

P

PACKAGING

This term is used for the design of packaging for objects as well as for their labeling and graphic design.

PICTOGRAM

A standardized, schematic drawing which can be immediately understood and indicates a direction, warns of a danger, or gives specific information.

PLASTIC

From Bakelite, the first synthetic plastic material, which appeared in 1907, to the range of over two hundred kinds of plastic today, plastic is the top design material. These molecules created by chemists from petrol take on all manner of shapes imagined by designers, according to processes of transformation and reworking specific to each kind of plastic.

PLYWOOD

Generally used in the manufacturing of designer furniture, this material appeared in the early twentieth century. It takes the form of panels composed of several thin layers of pressurized wood stuck together. Flexible and resistant.

POSTMODERNISM

Modernism was marked in the field of applied arts by the joining together of art and industry which gave rise,

Eau de Paris, Pierre Charpin, 2004.

since the 1920s, to a unprecedented stylistic rupture and triggered the reign of rational forms and new materials. In the late 1970s, designers from the Italian group Alchimia questioned the formal austerity of functionalism and defended the return of fantasy and a baroque attitude. Their rebellious attitude emphasized the transition from modernism to postmodernism in design. It chimes with the new industrial revolution represented by Internet and the global market.

PROCESS

A chain of pre-established tasks which respond to a precise plan in view of a determined result. In design, numerous consumer products are made according to this production method.

PROSPECTIVE

Prospective is industrial design's favorite field of research. It supplies the development strategies of many firms (in particular automobile manufacturers). It is based on various subjects: sociology, marketing, fashion, technology, etc. Teams then elaborate scenarios that generate innovative concepts and designs.

PROTOTYPE

Original copy for the experimentation of a new product in view of mass production.

R

REPRODUCTIONS

Since the 1980s, small and medium-sized companies have begun to reproduce design classics in limited editions. For example, Cassina for Le Corbusier, Rietveld, and Chareau; Vitra for Eames; and Ecart International for Fortuny, Gray, and Mallet-Stevens.

S

SERIES

A product formatted and manufactured in great numbers according to methods which allow for a low unit price. We talk of limited editions when models are manufactured in a small quantity which signifies rarity: and so prices rise!

SPECIFICATIONS

A list of specifications submitted to the designer when he responds to a commission. He must, on the one hand, develop the firm's goals and codes, and on the other, take into account the moral and environmental parameters at stake.

T

TYPOGRAPHY

Layout of texts to be printed. Creation of characters. Techniques and processes for the printing of texts through the mounting of characters in relief.

V

VISUAL IDENTITY

Everything that constitutes the image of a brand is used as a means of communication: logo, packaging, labels, vehicles, commercials, etc. Metro Goldwyn Mayer's roaring lion or the topography of the Monoprix name function as reference points.

Digestion collection, wall shelf, polyethylene foam, strap, and aluminum, Matali Crasset, Edra, 1998–2000.

Papelino-Bell,
bell in colored
Murano crystal,
(traditional Venetian
glass), Campana
brothers, Campane
di Campana
collection, 2005.

WHERE TO SEE DESIGN?

You don't have to go to a gallery or a museum to see design—it's everywhere: at home, in the office, or in public spaces. Some places are devoted to design, which function as display windows showcasing the latest fashions. These are often hotels, restaurants, or shopping centers created by great names in design.

MUSEUMS, INSTITUTIONS

AUSTRALIA

Powerhouse Museum
500 Harris Street Ultimo
P.O. Box K346 Haymarket
Sydney NSW 1238
Australia
+61 2 9217 0111
www.powerhousemuseum.
com/

BELGIUM

Grand-Hornu
Rue Sainte-Louise 82
7301 Hornu
+32 65 65 21 21
www.grand-hornu.be

CANADA

Centre de Design
Université du Quebec
à Montréal
1440 rue Sanguinet
Montréal H2X 3X9
+1 514 987 3395
www.centrededesign.com/
centre.html

Design Exchange
234 Bay Street
P.O. Box 18
Toronto Dominion Centre
Toronto, ON M5K 1B2
+1 416 363 6121
www.dx.org

DENMARK

**The Danish Museum
of Decorative Art**
Bredgade 68
1260 Copenhagen K
+45 33 18 56 56
www.kunstindustrimuseet.
dk

FRANCE

**Arc en Rêve, Centre
d'Architecture**
7 rue Ferrère
33000 Bordeaux
+33 5 56 52 78 36
www.arcenreve.com

Les Arts Décoratifs
107 rue de Rivoli
75001 Paris
+33 1 44 55 57 50
www.ucad.fr

Centre Pompidou
Place Beaubourg
75004 Paris
+33 1 44 78 12 33
www.centrepompidou.fr

**Centre de Recherche
sur les Arts du Feu et
de la Terre (CRAFT)**
Impasse Becquerel
87000 Limoges
+33 5 55 49 17 17
www.craft-limoges.org

Centre de Recherche
sur le Verre et les Arts
Plastiques (CIRVA)
62 rue de la Joliette
13002 Marseille
+33 4 91 56 11 50

Fondation Cartier
261 boulevard Raspail
75014 Paris
+33 1 42 18 56 50
www.fondation.cartier.fr

**Fonds National d'Art
Contemporain**
www.fnac.culture.gouv.fr

GERMANY

Bauhaus Foundation
Gropiusallee 38
06846 Dessau
+49 340 6508 251
www.bauhaus-dessau.de

Pinakothek der Moderne
Barerstrasse 40
80333 Munich
+ 49 89 27 272 50
www.die-neue-
sammlung.de

Thonet Museum
Michael-Thonet Strasse 1
35059 Frankenberg/Eder
+49 645 1508 119
www.thonet.de

Vitra Design Museum
Charles-Eames-Str. 1
D-79576 Weil am Rhein
+ 49 7621 702 3200
www.design-museum.de

ITALY

Triennale di Milano
Viale Alemagna 6
20121 Milano
+39 027 24 341
www.triennale.it

SOUTH AFRICA

The Design Museum
The Foundry
Ebenezer Road
Green Point
Cape Town
+27 21 418 1154
info@designmuseum.co.za

UNITED KINGDOM

Design Museum
Shad Thames
London SE1 2YD
+44 870 909 9009
www.designmuseum.org

UNITED STATES

A+D
6032 Wilshire Blvd
Los Angeles, CA 90036
+1 323 932 9393
www.aplusd.org/

The Chicago Athenaum International Museum of Architecture and Design
307 N. Michigan Avenue
Chicago, IL 60601
+1 312 372 1083
www.chi-athenaeum.org

Cooper-Hewitt National Design Museum
2 East 91st Street
New York, NY 10128
+1 212 849 8400
ndm.si.edu/

The Eisner: American Museum of Advertising & Design
208 North Water Street
Milwaukee, WI 53202
USA
+ 1 414 847 3290
www.eisnermuseum.org/

Museum of Arts and Design (MAD)
2 Columbus Circle
New York, NY 10019
+ 1 212 299 7777
www.madmuseum.org/

Museum of California Design
P.O. Box 361370
Los Angeles, CA 90036
+1 323 930 2700
www.mocad.org/

Museum of Design Atlanta
285 Peachtree Center Avenue
Marquis II Tower
Atlanta, Georgia 30303
+1 404 979 6455
www.museumofdesign.org/

Museum of Modern Art (MoMA)
11th West 53rd Street
New York, NY 10019-5497
+1 212 708 9400
www.moma.org

National Ornamental Metal Museum
374 Metal Museum Drive
Memphis, TN 38106
+1 901 774 6380 or
+1 877 881 2326 (toll free)
www.metalmuseum.org/

San Francisco Museum of Crafts Design
550 Sutter Street
San Francisco, CA 94102
+1 415 773 0303
www.sfmcd.org/

The Wolfsonian
1001 Washington Avenue
Miami Beach, FL 33139
+1 305 531 1001
www.wolfsonian.fiu.edu/

SALONS, FAIRS, EVENTS

CANADA
Festival of Architecture and Design (Toronto)
www.toronto.ca/fad/

FRANCE
Salon du Meuble (Paris)
www.salondumeuble.com

Maison et Objet (Paris)
www.maison-objet.com

Designer's Days (Paris)
www.designersdays.com

Les Puces du Design (Paris)
www.pucesdudesign.com

La Biennale Internationale de Design de Saint-Étienne
www.institutdesign.com

FINLAND
Helsinki Design Week
www.helsinkidesignweek.com/

GERMANY
International Design Festival Berlin
dmy-berlin.com/

ITALY
Salone Internationale del Mobile (Milan)
www.cosmit.it

UNITED KINGDOM
100% Design (London)
www.100percentdesign.co.uk

Design Event (Newcastle)
www.design-event.co.uk/

The London Design Festival
www.londondesignfestival.com/

UNITED STATES
Design Miami
www.designmiami.com

Design Philadelphia
designphiladelphia.org/

International Contemporary Furniture Fair (New York)
www.icff.com

National Design Week (Nationwide)
www.nationaldesignawards.org

Portland Architecture and Design Festival
www.aiaportland.org

Whippet Bench,
Radi designers,
1998.

SELECTED BIBLIOGRAPHY

GENERAL AND REFERENCE BOOKS

AMIC, YOLANDE. *Le Mobilier français 1945–1964.* Paris: Éditions du Regard, 1983.

ANARGYROS, SOPHIE. *Le Mobilier français 1980.* Paris: Éditions du Regard, 1984.

AUSSEL, ANDRE, AND CHARLES BARJONET. *Étude des styles du mobilier.* Rev. ed. Paris: Dunod, 1996.

BARTHES, ROLAND. *Mythologies.* Translated by Annette Lavers. London: J. Cape, 1972.

BAUDRILLARD, JEAN. *For a Critique of the Political Economy of the Sign.* Saint Louis: Telos Press, 1981.

BAUDRILLARD, JEAN. *The Consumer Society: Myths and Structures.* London: Sage, 1998.

BAUDRILLARD, JEAN. *The System of Objects.* London: Verso, 2006.

BENJAMIN, WALTER. *The Work of Art in the Age of its Technological Reproducibility and Other Writings on Media.* Cambridge, MA: Belknap Press, 2008.

DE BURE, GILLES. *Le Mobilier français 1965–1979.* Paris: Éditions du Regard, 1983.

CARIOU, JOËL. *Classiques du design.* Paris: Syros-Alternatives, 1988.

COLIN, CHRISTINE. *Design aujourd'hui.* Paris: Flammarion, 1988.

COLIN, CHRISTINE, ED. *Design and Utopias.* Paris: Industries françaises de l'ameublement, 2000.

COLIN, CHRISTINE, ED. *Design et Imitations.* Paris: Industries françaises de l'ameublement, 2004.

COLLIN DU BOCAGE, OLIVIER. *Reconnaître et choisir ses meubles.* Paris: Hachette Pratique, 2005.

DEFLASSIEUX, FRANÇOISE. *Guide des meubles et des styles.* Paris: Solar, 2005.

DE NOBLET, JOCELYN. *Design, le geste et le compas.* Paris: Somogy, 1988.

DE NOBLET, JOCELYN, ED. *Industrial Design: Reflection of a Century: 19th-21st Century.* Paris, Flammarion, 1996.

TROCHE, SUZANNE. *1900–2000: toute la décoration intérieure.* Paris: Octopus, 2003.

Design et sièges de collection. Paris: Industries Françaises de l'Ameublement, 2004.

Design français 1960–1990: trois décennies. Paris: APCI / Éditions du Centre Georges Pompidou, 1988.

Nouvelles tendances du design. Paris: Éditions du Centre Georges Pompidou, 1987.

Les Années Via 1980–1990. Paris: Union des Arts Décoratifs, 1990.

REVIEWS, MAGAZINES AND COLLECTIONS

Curve (Australia).

Domus (Italy).

Form (US).

Harvard Design Magazine (US).

ID Magazine (US).

Intramuros (France).

New Design (UK).

"Design." Special issue, *Beaux-Arts* (June 2003).

"Design et designers." Paris: Pyramyd Éditions. (monographs).

"Design-Monographs." Paris: Éditions Dis Voir.

"Designs that Changed the World." London: Conran.

"Dictionnaire du design." Paris: Editions du Seuil. (Germany, Great Britain, Italy, Scandinavia).

"MoMA Design Series." New York: The Museum of Modern Art.

SPECIALIST WORKS

BONY, ANNE. *Furniture and Interiors in the 1940s.* Paris: Flammarion, 2003.

BONY, ANNE. *Furniture and Interiors in the 1960s.* Paris: Flammarion, 2004.

DORMER, PETER. *Design since 1945.* New York: Thames & Hudson, 1993.

FAYOLLE, CLAIRE. *Le Design.* Rev. ed. Paris: Scala, 2005.

FRANCASTEL, PIERRE. *Art and Technology in the Nineteenth and Twentieth Centuries.* Translated by Randall Cherry. New York: Zone Books, 2003.

GUIDOT, RAYMOND. *Histoire du design 1940–2000.* Rev. ed. Paris: F. Hazan, 2000.

GUIDOT, RAYMOND. *Design, carrefour des arts.* Paris: Flammarion, 2003.

GUIDOT, RAYMOND, ED. *Industrial Design, Techniques, and Materials.* Paris: Flammarion, 2006.

GUILLAUME, VALERIE, BENOIT HEILBRUNN, AND OLIVIER PEYRICOT. *L'ABCdaire du design.* Paris: Flammarion, 2003.

HEILBRUNN, BENOIT. *Le Logo.* Rev. ed. Paris: PUF, 2006.

KLEIN, NAOMI. *No Logo: No Space, No Choice, No Jobs.* 10th anniv. ed. New York: Picador, 2009.

STEPHANE, LAURENT. *Chronologie du design.* Rev. ed. Paris: Flammarion, 2008.

SCHULMANN, DENIS. *Le Design industriel.* Paris: PUF, 1995.

TROY, NANCY J. *Modernism and the Decorative Arts in France: Art Nouveau to Le Corbusier.* New Haven and London: Yale University Press, 1991.

WHITFORD, FRANK. *Bauhaus.* London: Thames & Hudson, 1984.

INDEX
OF PROPER
NAMES

PHOTOGRAPHIC CREDITS

p. 1: © Stuart Haygarth ◦ **p. 2:** © 5.5 designers. ◦ **p. 3:** © Dyson ◦ **p. 4:** © Studio Campana/photo Luis Calazans ◦ **p. 5:** © Stefan Zwicky–Architekt BSA/SIA ◦ **p. 6:** © Pablo Reinoso ◦ **p. 7:** © Ron Arad ◦ **p. 8:** © Gaetano Pesce studio, New York ◦ **p. 14:** © Alessi ◦ **p. 15:** © Philippe Starck/Printemps/Moulin Roty ◦ **p. 16:** top: © Wilkinson Sword Limited/Pentagram/Kenneth Grange, bottom: Marc Newson Ltd/photo Patrick Gries ◦ **p. 17:** © Artek ◦ p. 19: © Centre Pompidou–Bibliothèque Kandinsky/ADAGP Paris 2006 ◦ **p. 20:** top: © Centre Pompidou–Bibliothèque Kandinsky, bottom: © FLC/ADAGP Paris 2006/Centre Pompidou–Bibliothèque Kandinsky. ◦ **p. 21:** © Rights reserved/photo CNAC/MNAM Dist. RMN–©Jacques Faujour ◦ **p. 22:** top: © courtesy Galerie Air de Paris/Jean-François Moriceau and Pétra Mzryk, middle: © IBM, bottom: © Paris Match ◦ **p. 23** top: © Sony. This model is no longer on the market and serves only as an illustration, bottom: © Adidas ◦ **p. 24:** © Pierre Hermé/photo Paul Goirand ◦ **p. 25:** © Nabaztag by Violet www.violet.net ◦ **p. 27:** © Philippe Starck/photo Jean-Baptiste Mondino. ◦ **p. 28:** left: © Dyson, right: © Marc Newson Ltd/photo Patrick Gries ◦ **p. 29:** © Christian Ghion/Daum ◦ **p. 33:** © Baccarat ◦ **p. 35:** © Andrée Putman/photo Deidi Von Schaewen. ◦ **p. 37:** © Demakersvan/photo Raoul Kramer ◦ **p. 39:** © Atelier Mendini. ◦ **p. 41:** © VIA/photo J. Bello ◦ **p. 44:** © Société Bic ◦ **p. 45:** © Ora Ito Studio ◦ **p. 46:** © Marc Newson Ltd./Biomega/photo Lena Paaske. ◦ **p. 47:** © Tupperware ◦ **p. 48:** © Radi designers ◦ **p. 49:** © Citroën Communication/photo E.Gautrand ◦ **p. 50:** © Martin Szekely/Perrier. ◦ **p. 51:** © Marc Newson Ltd./Dom Pérignon ◦ **p. 53:** © Hussein Chalayan/photo Chris Moore. ◦ **pp. 54–55:** © Miyake Design Studio/photo Pascal Roulin ◦ **p. 57:** © Courtesy Galerie Dominique Fiat/photo Patrick Gries ◦ **p. 58:** © Courtesy Galerie Kreo/photo Marc Domage ◦ **p. 59:** © Courtesy Lara Vincy/photo Olivier Liégent, 2006 ◦ **p. 61:** top: © Patrick Jouin/photo Eric Laignel, bottom: © JCDecaux ◦ **p. 62:** © Andrée Putman/photo Christophe Kicherer ◦ **p. 63:** © Ingo Maurer and team 2005/photo Nacasa & Partners, Tokyo ◦ **p. 66:** © Red and Blue–Gerrit T. Rietveld–"Cassina I Maestri" Collection/photo Mario Carrieri ◦ **p. 67:** top and middle: © Galerie Vallois – Paris, left, bottom: © Musée des années 30, Boulogne-Billancourt/photo Philippe Fuzeau, right, bottom: © www.ecart-international.com/photo Philippe Costes. ◦ **pp. 68–69:** © photo Nicolas Matheus for *Elle Décoration* ◦ **p. 70:** © Musée des années 30, Boulogne-Billancourt/photo Arnaud Olszak ◦ **p. 71:** top: © www.ecart-international.com/photo Philippe Costes, middle: © Bang and Olufsen, left, bottom: © photo Pascal Cadiou/Musée des années 30, Boulogne-Billancourt, right, bottom: © FLC/ADAGP L1(20)3 Paris 2006 ◦ **pp. 72–73:** © photo Marianne Haas for *Elle Décoration* ◦ **p. 74:** © photo CNAC/MNAM Dist. RMN–©Jean-Claude Planchet ◦ **p. 75:** top: © photo CNAC/MNAM Dist. RMN–©Jean-Claude Planchet, left, middle: © photo CNAC/MNAM Dist. RMN–©Jacques Faujour, right, middle: © Kenwood Manufacturing Company, bottom: © Courtesy Jousse Entreprise ◦ **p. 76:** © Knoll ◦ **p. 77:** left, top: © Courtesy Jousse Entreprise, right, top: © Kenwood manufacturing Company, top, middle: © Courtesy Conservatoire SEB, Lyon (69), right, middle: © Braun, bottom: © Courtesy Jousse Entreprise ◦ **p. 78:** © L'illustration ◦ Arne Jacobsen/Fritz Hansen ◦ **p. 79:** © Arne Jacobsen/Fritz Hansen/photo Strüwing. ◦ **p. 80:** © ADAGP Paris 2006/photo CNAC/MNAM Dist. RMN–©Jean-Claude Planchet ◦ **p. 81:** left, top: © Renault, right, top: © Courtesy Conservatoire SEB, Lyon (69), left, middle: © M.A Barjou/Fonds National d'art contemporain, right, middle: © Musée de la Ville de Saint-Quentin-en-Yvelines, photo J.-C. Bardot, bottom: © Collection Daniel Bouteiller ◦ **pp. 82–83:** © Collection du Mobilier National/photo Studio Contact in Beauvoir ◦ **p. 84:** © photo CNAC/MNAM Dist. RMN–© Georges Meguerditchian ◦ **p. 85:** top: © Olivetti, middle: © Sony. This model is no longer on the market and serves only as an illustration, left, middle: © Courtesy Gufram, Balangero (Italy)/Collection du Fonds national d'art contemporain/Cnap, Paris, right, bottom: © Collection du Fonds national d'art contemporain/Cnap, Paris/photo Michel-Alain Barjou, Paris. ◦ **p. 86:** © Verner Panton design. ◦ **p. 87:** © photo Pierre Berdoy/Collection du Mobilier National. ◦ **p. 88:** © Memphis Milan ◦ **p. 89:** right, top: © Swatch ◦ top, middle: © Musée d'Art Moderne, Saint-Etienne Métropole/photo Yves Bresson, right, top: © Alessi, middle: © Sylvain Dubuisson/Collection des arts déco/photo Laurent Sully-Jeaulmes, right, bottom: © Centre Pompidou–Bibliothèque Kandinsky, left, bottom: © Collection du Fonds national d'art contemporain/Cnap, Paris/photo V.Chenot, Paris ◦ **p. 90:** © Andrée Putman/photo Deidi Von Schaewen ◦ **p. 91:** © Philippe Starck/photo Deidi Von Schaewen ◦ **p. 92:** © Apple. ◦ **p. 93:** top: © Tsé & Tsé Associées, bottom: © Bang and Olufsen, left, bottom: © Konstantin Gcric Industrial Design, right, bottom: © Marc Newson/Collection du Fonds national d'art contemporain/Cnap, Paris/photo Bruno Scotti. ◦ **p. 94:** © Godefroy de Virieu ◦ **p. 95:** top © Courtesy of Carpenters Workshop Gallery, London, middle: Courtesy agence 14 septembre/Edra, © Apple left, bottom: © Courtesy Arik Levy/ENO, right, bottom: © Philippe Starck/Courtesy Flos ◦ **p. 101:** © Sentou Edition ◦ **p. 102:** © Marc Newson Ltd/Pod. ◦ **p. 103:** © Collection du Fonds national d'art contemporain/Cnap, Paris/photo Michel-Alain Barjou ◦ **p. 105:** top: © Courtesy Avant-Scène, bottom: © Pucci de Rossi ◦ **pp. 106–107:** © Moroso ◦ **p. 108:** © Morrison Studio, London/Alessi. ◦ **p. 109:** © Ronan and Erwan Bouroullec/photo Morgan Le Gall ◦ **p. 110:** © Art Digital Studio/Communic'art 2006 ◦ **p. 111:** © Zig Zag–Gerrit T. Rietveld––"Cassina I Maestri" Collection/Cassina/photo Mario Carrieri ◦ **p. 113:** top: © David Trubridge/distribution MOA, bottom: © Marcel Wanders ◦ **p. 114:** © Ronan Bouroullec ◦ **p. 115:** © Studio Jurgen Bey/Droog Design for Oranienbaum photo Marsel Loermans ◦ **p. 118:** © collection du Fonds national d'art contemporain/Cnap, Paris/édition Altras ◦ **p. 119:** © collection du Fonds national d'art contemporain/Cnap, Paris ◦ **p. 120:** © Zanotta Spa/photo Marino Ramazzoti ◦ **p. 121:** © Designfenzider Studio ◦ **p. 123:** © 5.5 designers ◦ **p. 124:** © Edra ◦ **p. 125:** © Stuart Haygarth ◦ **p. 127:** © JongeriusLab/Courtesy Galerie Kreo ◦ **p. 128:** © édition Neotu/Galerie Mouvements modernes/photo Deidi Von Schaewen. ◦ **p. 129:** © Designfenzider Studio ◦ **p. 131:** left: © Jurgen Bey studio, right: © Maarten Baas ◦ **p. 132:** © Pablo Reinoso ◦ **p. 133:** © Stefan Zwicky–Architekt BSA/SIA ◦ **p. 135:** © photo Vincent Thibert for *Elle Décoration* ◦ **pp. 136–137:** © Studio Jurgen Bey/assisted by Henriette Waal/photo Bob Goedewaagen ◦ **p. 140:** © Freeplay ◦ **p. 141:** top: © Jasper Morrison Ltd/Vitra/photo André Huber ◦ **p. 143:** © Inflate ◦ **p. 144:** © Du Bonheur!/photo Gilbert Lamurseau. ◦ **p. 145:** © www.azuamoline.com/photo Daniel Riera ◦ **p. 147:** © Igor Novitzki & Yann Guidon/Courtesy Galerie numeriscausa. ◦ **p. 148:** © NEC Corporation ◦ **p. 149:** © Frédéric Ruyant/photo Christophe Fillioux ◦ **p. 150:** © édition Kréo-Radi designers/Collection Frac Nord Pas-de-Calais (Fond Régional d'art contemporain)/photo S. Garcia ◦ **p. 151:** © Montina ◦ **p. 155:** © Michel Mallard ◦ **p. 156:** left: © Centre Pompidou–Bibliothèque Kandinsky/Thonet, right: © Knoll ◦ **p. 157:** left: © Thonet GmbH, Germany, right: © Artek ◦ **p. 158:** left: © Knoll, right: © Vitra/photo Hans Hansen ◦ **p. 159:** left: © Licensed by ABV/Tecno SpA/photo Pietro Carrieri, right: © Knoll ◦ **p. 160:** left: © Artifort, right: © courtesy Fiell International Ltd., London ◦ **p. 161:** left: © Kartell, right: © Menzdit-Werke, Menzingen/Die Neue Sammlung/Pinakothek der Moderne, Munich, Germany ◦ **p. 162:** left and right: Zanotta Spa/photo Marino Ramazzoti ◦ **p. 163:** left: © Yves Chenot, Paris/collection du Fonds national d'art contemporain/Cnap, Paris/Verner Panton design. ◦ **p. 163:** right: Courtesy B&B Italia,

254

acknowledgments

The author would like to thank her close friends and family for their support and encouragement, in particular Félix and Pierre Desfons, Catherine Schwaab and Marie-Paule Serre, Anne Bony, and Élisabeth Vedrenne.

She thanks her editor Laurence Basset for her trust and determination to bring this new project to life.

She also thanks her collaborator Nicolas Rabeau for his precious advice, François Huertas for his talented graphic design, Marguerite Vial for her precise research, and Michelle Teboul for her efficiency.

The author and editor would like to extend sincere thanks to the designers, manufacturers, photographers, and gallery owners who have generously offered their publication rights for a better understanding of design, as well as to all those who lent their photographic documents.

Translated from the French by ANNA HIDDLESTON
Design: FRANÇOIS HUERTAS
Photo Research and Documentation: MARGUERITE VIAL
Copyediting: HELEN WOODHALL
Typesetting: THIERRY RENARD
Proofreading: CHRISOULA PETRIDIS
Color Separation: REPROSCAN
Printed in Slovenia by KOROTAN

Café Marly collection, Olivier Gagnère, Artelano, 1994.

Distributed in North America by Rizzoli International Publications, Inc.

Originally published in French as *Le design hier, aujourd'hui, demain: mode d'emploi* by HFA/Editions Filipacchi, 2006
Revised French edition
© Flammarion, S.A., Paris, 2009

English-language edition
© Flammarion, S.A., Paris, 2010

editions.flammarion.com

10 11 12 3 2 1

ISBN: 978-2-08-030132-1

Dépôt légal: 03/2010